S0-DTD-130

Dictionary of Food Ingredients

Third Edition

Dictionary of Food Ingredients

Third Edition

Robert S. Igoe
Kelco, A Division of Monsanto Company

& Y. H. Hui
Technology Commerce International

CHAPMAN & HALL

New York • Albany • Bonn • Boston • Cincinnati • Detroit • London • Madrid • Melbourne
Mexico City • Pacific Grove • Paris • San Francisco • Singapore • Tokyo • Toronto • Washington

Printed in the United States of America

For more information, contact:

Chapman & Hall
115 Fifth Avenue
New York, NY 10003

Chapman & Hall
2-6 Boundary Row
London SE1 8HN
England

Thomas Nelson Australia
102 Dodds Street
South Melbourne, 3205
Victoria, Austrailia

Chapman & Hall GmbH
Postfach 100 263
D-69442 Weinheim
Germany

Nelson Canada
1120 Birchmount Road
Scarborough, Ontario
Canada M1K 5G4

International Thomson Publishing Asia
221 Henderson Road #05-10
Henderson Building
Singapore 0315

International Thomson Editores
Campos Eliseos 385, Piso 7
Col. Polanco
11560 Mexico D. F. Mexico

International Thomson Publishing - Japan
Hirakawacho-cho Kyowa Building, 3F
1-2-1 Hirakawacho-cho
Chiyoda-ku, 102 Tokyo
Japan

1 2 3 4 5 6 7 8 9 10 XXX 01 00 99 98 97 96 95

Library of Congress Cataloging-in-Publication Data

Igoe, Robert S.
Dictionary of food ingredigents / by Robert S. Igoe and Y. H. Hui.
-- 3rd ed.
p. cm.
Includes bibliographical references and index.
ISBN 0-412-07281-5 (cloth : alk. paper). -- ISBN 0-412-07291-2
(pbk. : alk. paper)
1. Food--Composition--Dictionaries. I Hui, Y. H. (Yiu H.)
II. Title.
TX551..T26 1995 95-4648
641' .034--dc20 CIP

British Library Cataloguing in Publication Data available

To order for this or any other Chapman & Hall book, please contact **International Thomson Publishing, 7625 Empire Drive, Florence, KY 41042.** Phone: (606) 525-6600 or 1-800-842-3636.
Fax: (606) 525-7778. e-mail: order@chaphall.com.

For a complete listing of Chapman & Hall's titles, send your request to
Chapman & Hall, Dept. BC, 115 Fifth Avenue, New York, NY 10003.

Contents

Preface

The *Dictionary of Food Ingredients* is a unique, easy-to-use source of information on over 1,000 food ingredients. Like the previous editions, the new and updated Third Edition provides clear and concise information on currently used additives, including natural ingredients, FDA-approved artificial ingredients, and compounds used in food processing. The dictionary entries, organized in alphabetical order, include information on ingredient functions, chemical properties, and uses in food products. The updated and revised Third Edition contains approximately 150 new entries, and includes an updated and expanded bibliography. It also lists food ingredients according to U. S. federal regulatory status.

Users of the two previous editions have commented favorably on the dictionary's straightforward and clearly-written definitions, and we have endeavored to maintain that standard in this new edition. We trust it will continue to be a valuable reference for the food scientist, food processor, food product developer, nutritionist, extension specialist, and student.

R. S. Igoe
Y. H. Hui

Ingredients

A

Acacia See **Arabic.**

Acesulfame-K A non-nutritive sweetener, also termed acesulfame potassium. It is a white, crystalline product that is 200 times sweeter than sucrose. It is not metabolized in the body. It is relatively stable as a powder and in liquids and solids which may be heated. Acesulfame-K is approved for use in dry food products.

Acesulfame Potassium See **Acesulfame-K.**

Acetanisole (p-methoxyacetophenone) A solid, pale yellow flavoring agent with a Hawthorn-like odor. It is soluble in most fixed oils and propylene glycol, and it is insoluble in glycerine and mineral oil. It is obtained by chemical synthesis. This flavoring substance or its adjuvant may be safely used in food in the minimum quantity required to produce its intended flavor. It can be used alone or in combination with other flavoring substances or adjuvants.

Acetic Acid An acid produced chemically from the conversion of alcohol to acetaldehyde to acetic acid. It is the principal component of vinegar which contains not less than 4 g of acetic acid in 100 cm^3 at 20° C. The approved salts include sodium acetate, calcium acetate, sodium diacetate, and calcium diacetate. It is used as a preservative, acidulant, and flavoring agent in catsup, mayonnaise, and pickles. It can be used in conjunction with leavening agents to release carbon dioxide from sodium bicarbonate.

Acetic Acid Ester of Monoglyceride See **Acetylated Monoglyceride.**

Acetic Acid, Glacial See **Glacial Acetic Acid.**

Acetic Anhydride An esterifier for food starch; also used in combination with adipic anhydride.

Acetone Peroxide A dough conditioner, maturing, and bleaching agent that is a mixture of monomeric and linear dimeric acetone peroxides which are strong oxidizing agents. It is used to age and bleach flour.

Acetylated Monoglyceride An emulsifier manufactured by the interesterification of edible fats with triacetin in the presence of catalysts or by the direct acetylation of edible monoglycerides with acetic anhydride without the use of catalysts. It is characterized by sharp melting points, stability to oxidative rancidity, film forming, stabilizing, and lubricating properties. It is used as a protective coating for meat products, nuts, and fruits to improve their appearance, texture, and shelf life. The coatings are applied by spraying, panning, and dipping. It is used in cake shortening and fats for whipped topping to enhance the aeration and improve foam stabilization. It is found in dry-mix whipped topping.

Acetylated Tartaric Acid Monoglyceride See **Diacetyl Tartaric Acid Esters of Mono- and Diglycerides.**

Acetyl Tartrate Mono- and Diglyceride See **Diacetyl Tartaric Acid Esters of Mono- and Diglycerides.**

Acid Calcium Phosphate See **Monocalcium Phosphate.**

Acid Casein The principal milk protein which is prepared from skim milk by precipitation with an acid, such as lactic, sulfuric, or hydrochloric acid, to lower the pH of the milk to 4.4 to 4.7. Caseins are identified according to the type of acid used, but in this form have little utility in foods, though they are used to some extent in cereal and bread fortification. Neutralization of the caseins yields the salts of which sodium and calcium caseinates are the most common. See **Casein.**

Acid-Modified Corn Starch See **Corn Starch, Acid-Modified.**

Acid Sodium Pyrophosphate See **Sodium Acid Pyrophosphate.**

Acidulants Acids used in processed foods for a variety of functions that enhance the food. They are used as flavoring agents, preservatives in microbial control, chelating agents, buffers, gelling and coagulating agents, and in many other ways.

Aconitic Acid A flavoring substance which occurs in the leaves and tubers of *Aconitum napellus L.* and other *Ranunculaceae.* Transaconitic acid can be isolated during sugar cane processing, by precipitation as the calcium salt from cane sugar or molasses. It may be synthesized by sulfuric acid dehydration of citric acid but not by the methanesulfonic acid method. It is used in a maximum level, as served, of 0.003 percent for baked goods, 0.002 percent for alcoholic beverages, 0.0015 percent for frozen dairy products, 0.0035 percent for soft candy, and 0.0005 percent or less for all other food categories.

Acrolein This is used in the ether etherification of food starch up to 0.6 percent and for the esterification and etherification of food starch up to 0.3 percent with vinyl acetate up to 7.5 percent.

Additives See **Food Additives.**

Adipic Acid An acidulant and flavoring agent. It is characterized as stable, nonhygroscopic, and slightly soluble, with a water solubility of 1.9 g per 100 ml at 20°C. It has a pH of 2.86 at 0.6 percent usage level at 25°C. It is used in powdered drinks, beverages, gelatin desserts, lozenges, and canned vegetables. It is also used as a leavening acidulant in baking powder. It can be used as a buffering agent to maintain acidities within a range of pH 2.5 to 3.0. It is occasionally used in edible oils to prevent rancidity.

Adipic Anhydride An esterifier for food starch in combination with acetic anhydride.

Agar A gum obtained from red seaweeds of the genera *Gelidium, Gracilaria,* and *Eucheuma,* class Rhodophyceae. It is a mixture of the polysaccharides agarose and agaropectin. It is insoluble in cold water, slowly soluble in hot water, and soluble in boiling water, forming a gel upon cooling. The gels are characterized as being tough and brittle, setting at 32° to 40°C and melting at 95°C. A rigid, tough gel can be formed at 0.5 percent. Agar mainly functions in gel formation because of its range between melting and setting temperatures, being used in piping gels, glazes, icings, dental impression material, and microbiological plating. Typical use levels are 0.1 to 2.0 percent.

Agar-Agar See **Agar.**

Albumin Any of several water-soluble proteins that are coagulated by heat and are found in egg white, blood serum, and milk. Milk albumin is termed lactalbumin and milk albuminate and it contains 28 to 35 percent protein and 38 to 52 percent lactose. It is used as a binder in imitation sausage, soups, and stews.

Aldehyde C-16 See **Ethyl-Methyl-Phenyl-Glycidate.**

Aldehyde C-18 See **(Gamma)-Nonalactone.**

Algin Gum derived from alginic acid which is obtained from brown seaweed genera, such as *Macrocystis pyrifera.* The derivatives are sodium, ammonium, and potassium alginates of which the sodium salt is most common. They are used to provide thickening, gelling, and binding. A derivative designed for improved acid and calcium stability is propylene glycol alginate. The algins are soluble in cold water and form non-thermoreversible gels in reaction with calcium ions and under acidic conditions. Algin is used in ice cream, icings, puddings, dessert gels, and fabricated fruit.

Alginate A gum derived from alginic acid that is used to provide thickening, gelling, and binding. See **Algin.**

Alginic Acid The acidic, insoluble form of algin that is a white to yellowish fibrous powder obtained from the brown seaweed genera such as *Macrocystis pyrifera.* The derivatives are soluble and include sodium, potassium, and ammonium alginate and propylene glycol alginate. It is used as a tablet disintegrant and as an antacid ingredient.

All-Purpose Flour A flour that is intermediate between long-patent flours (bread flour) which contain more than 10.5 percent protein and 0.40 to 0.50 percent ash and short-patent flours (cake flour) which generally contain less than 10 percent protein and less than 0.40 percent ash. It is made from hard or soft wheat and is used in baking and in gravies. It is also termed family flour.

Allspice A spice made from the dried, nearly ripe berries of *Pimenta officinalis,* a tropical evergreen tree. It has an aroma and flavor resembling that of a blend of nutmeg, cinnamon, and cloves. For labeling purposes, allspice refers to the spice of Jamaican origin. It is used in fruit pies, cakes, mincemeat, plum pudding, soups, and sauces.

Allyl Anthranilate A synthetic flavoring agent that is a light yellow colored liquid of green leaf-wine odor. It is stable but may cause discoloration. It should be stored in glass or tin containers. It is used as flavoring for its wine note and has application in beverages and candy at 1 to 2 parts per million.

Allyl Caproate See **Allyl Hexanoate.**

Allyl Cinnamate A synthetic flavoring agent that is a fairly stable, hazy, colorless to light-yellow-colored liquid of cherry odor. It is used for its cherry note in flavors and has application in baked goods and candies at 1 to 2 parts per million.

Allyl-2,4-Hexadienoate See **Allyl Sorbate.**

Ally Hexanoate (allyl caproate) A liquid flavoring agent with a strong pineapple odor and pale-yellow color. It is practically insoluble in propylene glycol and miscible with alcohol, most fixed oils, and mineral oil. It is obtained by chemical synthesis. It can be used alone or in combination with other flavoring substances or adjuvants.

Allyl Isothiocyanate A synthetic flavoring agent that is a moderately stable, colorless to pale yellow liquid of pungent and irritating odor. It should be stored in glass containers. It is used as an artificial oil of mustard and as an imitation horseradish flavor with application in condiments, meats, and pickles at 87 parts per million. It is also termed mustard oil.

Allyl Isovalerate A synthetic flavoring agent that is a stable, colorless to light yellow liquid of fruity odor. It should b stored in glass or tin containers. It has usage in fruit flavors with applications in beverages, baked goods, ice cream, and candy at 8 to 50 parts per million.

Allyl Mercaptan A synthetic flavoring agent that is a stable, colorless liquid of garlic-like odor. It should be stored in glass or tin containers. It is used in artificial garlic flavors for application in condiments at 3 parts per million, in baked goods at 2 parts per million. It is also termed 2 propylene-1 thiol.

Allyl Nonanoate A synthetic flavoring agent that is a stable, colorless to light yellow liquid of fruity-cognac odor. It should be stored in glass or tin containers. It is used in fruit flavors like melon and pineapple for application in candy, ice cream, and beverages at 0.70 to 5 parts per million.

Allyl Octanoate A synthetic flavoring agent that is a colorless to light yellow liquid and has a fruity odor. It is alkali and mineral acid unstable and should be stored in glass, tin, and resin-lined containers. It is used to give flavors a fruity note and has application in dessert gels, puddings, beverages, and candy at 3 to 25 parts per million.

Allyl Phenoxyacetate A synthetic flavoring agent that is a stable, colorless to light yellow liquid of heavy fruit note odor. It should be stored in glass or tin containers. It is used in pineapple, quince, and fruit flavors with applications in candy and beverages at 1 to 3 parts per million.

Allyl Phenylacetate A synthetic flavoring agent that is a stable, colorless to light yellow liquid with a fruity odor of banana and honey. It should be stored in glass or tin containers. It is used in flavors for honey and has application in candy and baked goods at 10 to 15 parts per million.

Allyl Sorbate A synthetic flavoring agent that is a colorless to light yellow liquid of sharp fruity odor. It is subject to polymerization and should be stored in glass or tin containers. It is used in pineapple and other fruit flavors which have application in puddings, candy, and beverages at 1 to 2 parts per million. It is also termed allyl-2, 4-hexadienoate.

Almond A nut obtained from the almond tree *Prunus amygdalus.* It exists as a sweet or bitter variety, with the sweet variety being used as edible nuts and the bitter variety being used as a source of almond oil. The obtainable forms range from whole nuts to slices to powder. The nuts are used as snacks, as a garnish on pastries, and as a flavorant.

Almond Oil The oil of the bitter almond after the removal of hydrocyanic acid. It is a colorless to slightly yellow liquid having a strong principally almond-like aroma. It is used mainly in the pharmaceutical and cosmetic industry and also as a food flavoring agent.

Almond Paste A paste made by cooking sweet and bitter almonds which have been ground and blanched in combination with sugar. It consists approximately of two parts almond to one part sugar. It is used in pastries and cakes.

Alpha-Tocopherol See **Tocopherol.**

Alum A preservative, the inclusive term for several aluminum-type compounds such as aluminum sulfate and aluminum potassium sulfate. It is used with EDTA to prevent discoloration of potatoes and to maintain the firmness of shrimp packs. It is also used in pickles and pickle relish.

Aluminum Ammonium Sulfate A general purpose food additive that functions as a buffer and neutralizing agent. Its solubility is 1 g in 7 ml of water at 25°C and 1 g in 0.3 ml of boiling water. It is used in baking powders.

Aluminum Calcium Silicate An anticaking agent used in vanilla powder. It is also used in salt up to 2 percent.

Aluminum Nicotinate. The aluminum salt of nicotinic acid. It is a source of niacin in foods of special dietary use.

Aluminum Oleate The aluminum salt of oleic acid which is used as a binder, emulsifier, and anticaking agent. It is practically insoluble in water.

Aluminum Sodium Sulfate A general purpose food additive that functions as a buffer, neutralizing agent, and firming agent. It is anhydrous and slowly soluble in water. The dodecahydrate form is readily soluble in water. It is also termed soda alum.

Aluminum Sulfate A starch modifier and firming agent. The anhydrous form has a slow rate of solution while the hydrate form has a solubility of 1 g in approximately 2 ml of water and a 1% solution pH of approximately 3.5. It is used up to 2 percent in combination with not more than 2 percent of 1-octenyl succinic anhydride. It is used as a firming agent in pickle and vegetable processing and as a processing aid in baked goods, gelatins, and puddings.

Amidated Pectin The low-methoxyl pectin that results when some of the methoxyl groups are transformed into amide groups during deesterification with ammonia. These pectins function best in applications between 30 and 65 percent soluble solids content and pH 3.0 to 4.5. They usually require no more calcium ions than are already present in the fruit to obtain gelation. The gels formed are heat-reversible. Applications include flans and tart glazing. Also see **Pectin.**

Amino Acids The food additive amino acids may be safely used as nutrients added to foods. The food additive consists of one or more of the following individual amino acids in the free, hydrated, or anhydrous form or as the hydrochloride, sodium, or potassium salts: L-Alanine, L-Arginine, L-Asparagine, L-Asparatic acid, L-Cysteine, L-Cystine, L-Glutamic acid,

L-Glutamine, Aminoacetic acid (glycine), L-Histidine, L-Isoleucine, L-Leucine, L-Lysine, DL-Methionine (not for infant foods), L-Methionine, L-Phenylalanine, L-Proline, L-Serine, L-Threonine, L-Tryptophan, L-Tyrosine, or L-Valine. The additive(s) is used to significantly improve the biological quality of the total protein in a food containing naturally-occurring, primarily-intact protein that is considered a significant dietary protein source. The amount of the additive added for nutritive purposes plus the amount naturally present in free and combined (as protein) form should not exceed the levels of amino acids expressed as percent by weight of the total protein of the finished food.

Amioca See **Waxy Maize Starch.**

Ammoniated Glycyrrhizin See **Glycyrrhizin.**

Ammonium Alginate A gum that is the ammonium salt of alginic acid. It is cold water soluble and forms viscous solutions. It functions as a stabilizer and thickener and its uses include bakery fillings, gravies, and toppings.

Ammonium Bicarbonate A dough strengthener, a leavening agent, a pH control agent, and a texturizer. Prepared by reacting gaseous carbon dioxide with aqueous ammonia. Crystals of ammonium bicarbonate are precipitated from solution and subsequently washed and dried.

Ammonium Carbonate A dough strengthener, a leavening agent, a pH control agent, and a texturizer. It is prepared by the sublimation of a mixture of ammonium sulfate and calcium carbonate, and occurs as a white powder or a hard, white translucent mass.

Ammonium Caseinate The ammonium salt of casein. It has a high nutritional value and low sodium content and is used in foods and pharmaceuticals. See **Caseinates.**

Ammonium Chloride A dough conditioner and yeast food that exists as colorless crystals or white crystalline powder. Approximately 30 to 38 g dissolve in water at 25°C. The pH of a 1 percent solution at 25°C is 5.2. It is used as a dough strengthener and flavor enhancer in baked goods and as a nitrogen source for yeast fermentation. It is also used in condiments and relishes. Another term for the salt is ammonium muriate.

Ammonium Hydroxide An alkaline that is a clear, colorless solution of ammonia which is used as a leavening agent, a pH control agent, and a surface finishing agent. It is used in baked goods, cheese, puddings, processed fruits, and the production of caramels.

Ammonium Muriate See **Ammonium Chloride.**

Ammonium Persulfate A bleaching agent for food starch that is used at up to 0.075 percent and with sulfur dioxide up to 0.05 percent.

Ammonium Phosphate Dibasic A general purpose food additive that is readily soluble in water, with approximately 57 g dissolving in 100 g of water at 0°C. A 1 percent solution has a pH of 7.6 to 8.2. It is used as a dough strengthener, firming agent, leavening agent, and pH control agent. Its uses include baked goods, alcoholic beverages, condiments, and puddings. In bakery products up to 0.25 part per 100 parts by weight of flour is used.

Ammonium Phosphate Monobasic A general purpose food additive which is readily soluble in water. A 1 percent solution has a pH of 4.3 to 5.0. It is used as a dough strengthener and leavening agent in baked goods and as a firming agent and pH control agent in condiments and puddings. It is also used in baking powder with sodium bicarbonate and as a yeast food.

Ammonium Sulfate A dough conditioner, firming agent, and processing aid which is readily soluble in water with a solubility of approximately 70 g in 100 g of water at 0°C. The pH of a 0.1 molar solution in water is approximately 5.5. It is used in caramel production and as a source of nitrogen for yeast fermentation. In bakery products, up to 0.25 part per 100 parts by weight of flour is used.

Ammonium Sulfite An additive used in the production of caramel.

Amylcinnamaldehyde (amylcinnomaldehyde) A flavoring agent that is a liquid, yellow, with an odor similar to jasmine. It is insoluble in glycerine and propylene, soluble in fixed oils and mineral oil. It is obtained by chemical synthesis. It can be used alone or in combination with other flavoring substances or adjuvants.

Anhydrous Milkfat See **Butter Oil.**

Anise A spice that is the dried, ripe fruit of *Pimpinella anisum,* a small herb. The flavor is similar to fennel or licorice while the seed resembles caraway seed. It is used in beverages, soups, candy, liquors, and sweet pastries.

Anisyl Butyrate A synthetic flavoring agent that is a stable, colorless liquid of sweet cassic odor. It should be stored in glass or tin containers. It will intensify vanilla flavor and is used as a fixative. It is used in ice cream, candy, and baked goods at 5 to 15 parts per million.

Anisyl Formate A synthetic flavoring agent that is a fairly stable, colorless to light yellow liquid of floral odor. It should be stored in glass, tin, or resin-lined containers. It is used in berry flavors for applications in beverages, candy, and baked goods at 3 to 10 parts per million.

Anisyl Propionate A synthetic flavoring agent that is a stable, colorless liquid with a heliotrope odor. It should be stored in glass or tin containers. It is used in small concentrations to intensify vanilla, plum, and quince flavor for applications in beverages, baked goods, and candy at 6 to 20 parts per million.

Annatto A color source of yellowish to reddish-orange color obtained from the seed coating of the tree *Bixa orellanna.* The oil-soluble annatto consists mainly of bixin, a carotenoid soluble in fats and oils with the color which is produced being found in the fat portion of the food. It has a yellow hue, very good oxidation stability, fair light stability, and good heat stability but it is unstable above 125°C. The water-soluble annatto is norbixin (the product resulting when bixin is saponified and the methyl-ethyl group is split off) which is dissolved as a potassium salt in lye. It is readily soluble in aqueous alkalis with the coloring occurring in the protein and starch fraction of the food. It has a yellow to orange hue and precipitates in most acid foods. The usage level is 0.5 to 10 parts per million in the finished food. It is used in sausage casings, oleomargarine, shortening, and cheese.

Annatto Extract See **Bixin.**

Anticaking Agents and Free-Flow Agents Substances added to finely powdered or crystalline food products to prevent caking, lumping, or agglomeration. Agents include calcium silicate, iron ammonium citrate, silicon dioxide and yellow prussiate of soda.

Antimicrobial Agents See **Preservatives.**

Antioxidants Substances used to preserve food by retarding deterioration, rancidity, or discoloration due to oxidation. The most commonly used antioxidant formulations contain combinations of BHA (butylated hydroxyanisole), BHT (butylated hydroxytoluene), and propyl gallate. Natural antioxidants such as tocopherols and guaiac gum usually lack the potency of BHA, BHT, and propyl gallate combinations. Antioxidants are effective at low concentrations, that is, 0.02 percent or less.

Apple Vinegar See **Cider Vinegar.**

Arabic A gum obtained from breaks or wounds in the bark of *Acacia* trees. It dissolves in hot or cold water forming clear solutions which can be up to 50 percent gum acacia. The solubility in water increases with temperature. It is used in confectionary glazes to retard or prevent sugar crystallization and acts as an emulsifier to prevent fat from forming an oxidizable, greasy film. It functions as a flavor fixative in spray-drying to form a thin film around the flavor particle. It also functions as an emulsifier in flavor emulsions, as a cloud agent in beverages, and as a form stabilizer. It is also termed acacia.

Arabinogalactan A gum, being the plant extract obtained from larch trees. It is soluble in hot and cold water, the water solutions up to 60 percent being fluid and above 60 percent forming a thick paste. It is stable over a wide pH range and is relatively unaffected by electrolytes. Its limited uses include dressings and pudding mixes. It is also termed larch gum.

Arginine A nonessential amino acid that exists as white crystals or powder. It is soluble in water. It is used to improve the biological quality of the total protein in a food which contains naturally occurring primarily intact proteins and as a nutrient and dietary supplement.

Arrowroot A starch obtained from *Maranta arundinacea*, a perennial that produces starchy rhizomes. It is neutral in flavor and of clear color. It is used as a thickener, using one-third to one-half as much as the flour or cornstarch level. It is used in fruit sauce, pie fillings, and puddings.

Artificial Coloring See **Colors and Coloring Adjuncts.**

Artificial Flavors See **Flavoring Agents.**

Artificial Sweeteners See **Nonnutritive Sweeteners and Sweeteners.**

Ascorbic Acid It is termed vitamin C, a water-soluble vitamin that prevents scurvy, helps maintain the body's resistance to infection, and is essential for healthy bones and teeth. It is the most easily destroyed vitamin and processing is recommended in stainless steel or glass. Storage at below −18°C is recommended. In its dry form it is nonreactive, but in solution it readily reacts with atmospheric oxygen and other oxidizing agents. One part ascorbic acid is equivalent to one part erythorbic acid. It is used as a vitamin supplement in beverages, potato flakes, and breakfast foods; and as a dough conditioning agent to strengthen and condition bread roll doughs. It is also used as an antioxidant to increase shelf life in canned and frozen processed foods. It is used in conjunction with BHA, BHT, and propyl gallate to regenerate them following the chemical changes they undergo when they prevent fat rancidity in bologna and other meats. Other forms of ascorbic acid are isoascorbic (erythorbic) acid, sodium ascorbate, and sodium isoascorbate.

Ascorbyl Palmitate An antioxidant formed by combining ascorbic acid with palmitic acid. Ascorbic acid is not fat soluble but ascorbyl palmitate is, thus combining them produces a fat-soluble antioxidant. It exists as a white or yellowish white powder of citric-like odor. It is used as a preservative for natural oils, edible oils, colors, and other substances. It acts synergistically with alpha-tocopherol in oils/fats. It is used in peanut oil at a maximum level of 200 mg/kg individually or in combination.

Aspartame A synthetic sweetener that is a dipeptide, providing 4 calories per gram. It is synthesized by combining the methyl ester of phenylalanine with aspartic acid, forming the compound N-L-alpha-aspartyl-L-phenylalanine-1-methyl ester. It is approximately 200 times as sweet as sucrose and tastes similar to sugar. Its minimum solubility is at pH 5.2, its isoelectric point. Its maximum solubility is at pH 2.2. The solubility increases with temperature. Aspartame has a certain instability in liquid systems which results in a decrease in sweetness. It decomposes to aspartylphenylalanine or to diketopiperazine (DKP) and neither of these forms is sweet. The stability of aspartame is a function of time, temperature, pH, and water activity. It is not usually used in baked goods because it breaks down at the high baking temperatures. It contains phenylalanine, which restricts its use for those afflicted with phenylketonuria, the inability to metabolize phenylalanine. Approved applications include cold breakfast cereals, desserts, topping mixes, chewing gum, powdered beverages, and frozen desserts. The usage level ranges from 0.01 to 0.02 percent. It is also approved for use in soft drinks, where it is used as a replacement for saccharin or as a admixture to saccharin-containing diet soft drinks.

Aspartic Acid A nonessential amino acid that exists as colorless or white crystals of acid taste. It is slightly soluble in water. It functions to improve the biological quality of a total protein in a food containing naturally occurring primarily intact protein and as a nutrient and dietary supplement.

Azodicarbonamide A dough conditioner that exists as a yellow to orange-red crystalline powder practically insoluble in water. It is used in aging and bleaching cereal flour to produce a more manageable dough and a lighter, more voluminous loaf of bread. It is used in bread flours and bread as a dough conditioner. It can be used with the oxidizing agent potassium bromate. A typical use level is less than 45 parts per million.

B

Babassu Oil The oil obtained from the nut of the babassu palm, which is native to Brazil. It is similar to coconut oil and acts as a substitute, being used in vegetable fat-based products.

Baking Powder A leavening agent that consists of a mixture of sodium bicarbonate, one or more leavening agents such as sodium aluminum phosphate or monocalcium phosphate, and an inert material such as starch. The inert material keeps the reactive components physically separated and minimizes premature reaction. It should yield not less than 12 percent of available carbon dioxide.

Baking Soda Bicarbonate of soda, chemically known as sodium bicarbonate. It liberates carbon dioxide, a leavening gas, when heated or mixed with an acid. It is a component of baking powder and is used as a leavening agent.

Baker's Yeast Extract A flavoring agent resulting from concentration of the solubles of mechanically ruptured cells of a selected strain of yeast, *Saccharomyces cerevisiae.* It may be concentrated or dried. It is used at a level not to exceed 5 percent in food.

Baker's Yeast Glycan The dried cell walls of yeast, *Saccharomyces cerevisiae,* obtained from brewing. Bakers' yeast glycan is used as an emulsifier and thickener in salad dressing.

Baker's Yeast Supplement A nutrient supplement which is the insoluble proteinaceous material remaining after the mechanical rupture of yeast cells of *Saccharomyces cerevisiae* and removal of whole cell walls by centrifugation and separation of soluble cellular materials.

Balsam Peru Oil A flavoring agent, which is liquid, and yellow to pale green in color. It is viscous and has a has a sweet balsamic odor. It is insoluble in glycerin, slightly soluble in propylene glycol, soluble and turbid in mineral oil, and soluble in fixed oils. It is obtained by extraction or distillation of Peruvian Balsam obtained from *Myroxylon pereirae* Royal Klotsche of the *Leguminosae* family. It can be used alone or in combination with other flavoring substances or adjuvants.

Barley A cereal grain of which there are winter and spring types. It is used in malting (the conversion of grain to malt used in beer production) as malted barley. Malt flour is used in baking, cereals, and sauces. Pearled barley, in which the hull and outer kernel are removed by abrasive action, is found in barley soups. Barley flour and flakes are used in baked products. Barley is high in carbohydrates and contains protein, calcium, phosphorus, and B vitamins.

Barley, Malted See **Malted Barley.**

Basil (Sweet Basil) A spice obtained from the dried leaves and tender stems of *Ocimum basilicum L.* The fresh basil resembles licorice in flavor and the dried leaves have a lemony anise-like quality. This delicate herb can be used generously and has an affinity for tomato-based products. It is used in tomato based recipes, with vegetables, and in tomato sauces.

Bay Leaves A spice flavoring that consists of the dried leaves obtained from the evergreen tree *Laurus nobilis,* also called sweet bay or laurel tree. They have a sweet, herbaceous flavor and are used as an herb. They are aromatic when crushed and find use in meat, soup, and stew.

Beeswax The purified wax obtained from the honeycomb of the bee is insoluble in water and is sparingly insoluble in cold alcohol. It is used to glaze candy in chewing gum, in confections, and as a flavoring agent.

Beet Extract A natural red colorant obtained from beets that is of very good water solubility and has fair pH stability, poor heat stability, and good light stability. It is colored by betacyanins which include red and yellow compounds, the major red pigment being betanin. The betanin accounts for 75 to 95 percent of the total pigment content. It is available in concentrate and powder forms and is used in yogurt, beverages, candies, and desserts.

Beets, Dehydrated See **Beet Extract.**

Beet Sugar See **Sugar.**

Benne See **Sesame Seed.**

Bentonite A general purpose additive that is used as a pigment and colorant and to clarify and stabilize wine.

Benzaldehyde (Benzoid Aldehyde) A flavoring agent which is liquid and colorless, and has an almond-like odor. It has a hot (burning) taste. It is oxidized to benzoic acid when exposed to air and deteriorates under light. It is miscible in volatile oils, fixed oils, ether, and alcohol; it is sparingly soluble in water. It is obtained by chemical synthesis and by natural occurrence in oils of bitter almond, peach, and apricot kernel.

Benzoate of Soda See **Benzoic Acid.**

Benzoic Acid A preservative that occurs naturally in some foods such as cranberries, prunes, and cinnamon. It is most often used in the form of sodium benzoate because of the low aqueous solubility of the free acid. Sodium benzoate is 180 times as soluble in water at 25°C as benzoic acid. The salt in solution is converted to the acid which is the active form. The optimum pH range for microbial inhibition is pH 2.5 to 4.0. It is used in acid foods such as carbonated beverages, fruit juice, and pickles. It is also termed benzoate of soda.

Benzoic Aldehyde See **Benzaldehyde.**

Benzoyl Peroxide A colorless, crystalline solid with a faint odor of benzaldehyde resulting from the interaction of benzoyl chloride and a cooled sodium peroxide solution. It is insoluble in water. It is used in specified cheeses at 0.0002 percent of milk level. It is used for the bleaching of flour, slowly decomposing to exert its full bleaching action, which results in whiter flour and bread.

Benzyl Propionate A flavoring agent which is liquid, colorless and has a sweet, fruity odor. It is soluble in most fixed oils and alcohol, slightly soluble in propylene glycol, and insoluble in glycerin.

Beta-Apo-8′-Carotenal A colorant that is a carotenoid producing a light to dark orange hue. It has fair light stability, poor oxidation stability, and good pH stability. It is insoluble in water but is available in water-dispersible, oil-dispersible, and oil-soluble forms. It has vitamin A activity. It is used in cheese, and cheese sauces, and dressings. The maximum usage level is 33 parts per million. Related colorants are beta-carotene and canthaxanthin.

Beta-Carotene A colorant that is a carotenoid producing a yellow to orange hue. It has good tinctorial strength, fair light stability, poor oxidation stability, and good pH stability. It is insoluble in water but is available as water-dispersible, oil-dispersible, and oil-soluble forms. It has vitamin A activity. It has a natural resistance to ascorbic acid reduction in beverages and thus is used in orange-colored liquid products. It is used in margarine, oils, cheese, and puddings at levels required to produce the desired color. Related colorants are canthaxanthin and beta-apo-8′-carotenal.

BHA See **Butylated Hydroxyanisole.**

BHT See **Butylated Hydroxytoluene.**

Bicarbonate of Soda See **Sodium Bicarbonate.**

Biotin A water-soluble vitamin that is a nutrient and dietary supplement. It is relatively stable to heat and storage and is found in eggs, liver, peanuts, milk, and meat. It functions in the metabolism of carbohydrates, proteins, and fats. It is essential for the activity of many enzyme systems.

Birch An artificial flavoring used in soft drinks such as birch beer.

Bixin A carotenoid that is the main coloring component of annatto. It is obtained from the *Bixa orellana* tree. Bixin is soluble in fats and oils and the produced color is found in the fat fraction of the food. It has a yellow hue, very good oxidation stability, fair light stability, and good heat stability, but it is poor at very high temperatures, such as above 125°C. One part bixin is equivalent to 1.5 parts carotene. It is used at 0.5 to 10 parts per million in finished foods, such as margarine, salad dressings, popcorn oil, and baked goods. It is also termed annatto extract. See **Annatto.**

Bleached Flour Flour that has been whitened by the removal of the yellow pigment. The bleaching can be obtained during the natural aging of the flour or can be accelerated by chemicals that are usually oxidizing agents which oxidize the carotenoid pigments to a nearly colorless product. The oxidizing agents also improve the flour performance by their effect on the protein. The process improves the baking quality by allowing the formation of high ratio cakes that would be likely to collapse if prepared with untreated flour.

Bleaching Agents See **Flour Treating Agents.**

Bodying Agents See **Stabilizers and Thickeners.**

Bran The seed husks or outer coatings of cereals, such as wheat, rye, and oats, that are separated from the flour. It is used in bran flakes cereal.

Bread Flour A hard-wheat flour, which generally contains in excess of 10.5 percent protein and is obtained from straight or long patent flours. These flours have high absorption and good mixing tolerance which make them suitable in yeast-leavened breads.

Brilliant Blue FCF See **FD&C Blue #1.**

Bromated Flour A white flour to which potassium bromate is added at a level not to exceed 50 parts per million. It is used in baked goods.

Brominated Vegetable Oil (BVO) A vegetable oil whose density has been increased to that of water by combination with bromine. Flavoring oils are dissolved in the brominated oil which can then be added to fruit drinks. The adjustment of the specific gravity makes it possible to obtain stable finished beverages. If the oil phase gravity is too low the emulsion will form a ring and if it is too high a white precipitate may form. It is also used in formulating cloud agents. Its use is limited to 15 parts per million.

Brown Sugar A sweetener that consists of sucrose crystals covered with a film of cane molasses. Molasses gives it the characteristic color and flavor. There are three grades, light, medium, and dark, which vary in sucrose content and color. It is used in baked goods, glazes, toppings, and fillings.

Bulgur A precooked cracked wheat that retains the bran and germ fraction of the grain. It resembles whole wheat nutritionally and is sometimes termed parboiled wheat. It is an excellent source of whole grain, protein, and carbohydrates. It is reconstituted by cooking or soaking in liquid. It can be used in bread, casseroles, and salads, or can be eaten as such.

Bulking Agents See **Stabilizers and Thickeners.**

Butter, Clarified Butter that has undergone purification by the removal of solid particles or impurities that may affect the color, odor, or taste.

Butter Fat. See **Milkfat.**

Buttermilk The product that remains when fat is removed from milk or cream in the process of churning into butter. Cultured buttermilk is prepared by souring buttermilk, or more commonly skim milk, with a suitable culture that produces a desirable taste and aroma. It is used as a beverage, as an ingredient in baked goods, and in dressings.

Buttermilk, Dried The powder form of buttermilk. It is similar in composition to nonfat dry milk but of higher fat concentration, much of which is phospholipids which provide good emulsifying and whipping properties. It is used in dry mix, desserts, soups, and sauces.

Butter Oil The clarified fat portion of milk, cream, or butter obtained by the removal of the nonfat constituent. It contains not less than 99.7 percent milk fat, not more than 0.2 percent moisture, and not more than 0.05 percent milk solids nonfat. It is used in frozen desserts, puddings, and syrups. It is also termed anhydrous milkfat, or ghee.

Butyl Acetate (*n*-Butle Acetate) A flavoring agent which is a clear, colorless liquid possessing a fruity and strong odor. It is sparingly soluble in water and miscible in alcohol, ether, and propylene glycol.

Butyl Butyryllactate A synthetic flavoring agent that is a stable, colorless to light yellow liquid with the odor of cooked butter. It is miscible with alcohol and most fixed oils, soluble in propylene glycol, and insoluble in glycerine and water. It should be stored in glass, tin, or resin-lined containers. It is used in butter flavors with applications in baked goods, and candy at 14 to 60 parts per million.

Butyl Heptanoate A synthetic flavoring agent that is a stable, colorless liquid of fruity odor. It is stored in glass or tin containers. It is used in flavors such as apple, blackberry, and ginger beer with applications in candy and baked goods at 25 parts per million.

Butylated Hydroxyanisole (BHA) An antioxidant that imparts stability to fats and oils and should be added before oxidation has started. It is a mixture of 3-tert-butyl-4-hydroxyanisole and 2-tert-butyl-4-hydroxyanisole. In direct addition, the fat or oil is heated to 60 to 70°C and the BHA is added slowly under vigorous agitation. The maximum concentration is 0.02 percent based on the weight of the fat or oil. It may protect the fat-soluble vitamins A, D, and E. It is used singly or in combination with other antioxidants. It is used in cereals, edible fat, vegetable oil, confectionary products, and rice.

Butylated Hydroxytoluene (BHT) An antioxidant that functions similarly to butylated hydroxyanisole (BHA) but is less stable at high temperatures. It is also designated 2,6-di-tert-butyl-para-cresol. See **Butylated Hydroxyanisole.**

Butylhydroquinone See **Tertiary Butylhydroquinone.**

Butylparaben See **Parabens.**

Butyric Acid A fatty acid that is commonly obtained from butter fat. It has an objectionable odor which limits its use as a food acidulant or antimycotic. It is an important chemical reactant in the manufacture of synthetic flavoring, shortening, and other edible food additives. In butter fat, the liberation of butyric acid which occurs during hydrolytic rancidity makes the butter fat unusable. It is used in soy-milk-type drinks and candies.

C

Cacao Butter See **Cocoa Butter.**

Caffeine A white powder or needles that are odorless and have a bitter taste. It occurs naturally in tea leaves, coffee, cocoa, and cola nuts. It is a food additive used in soft drinks for its mildly stimulating effect and distinctive taste note. It is used in cola-type beverages and is optional in other soft drinks up to 0.02 percent.

Cake Flour A soft wheat flour that is generally a short patented flour containing less than 10 percent protein. Such flours are low in water absorption and are of short mixing time and tolerance. It is used in chemically leavened cakes, cookies, and pastries.

Calciferol A fat-soluble vitamin, termed vitamin D_2, which is stable unless oxidized. It is necessary for growth and maintenance of teeth and bones and the normal utilization of calcium and phosphorus; it is used medicinally in the treatment of rickets and as a dietary supplement. Its sources include fish liver, and vitamin D fortified milk.

Calcium An alkaline earth element that contributes toward bone and teeth formation, muscle contraction, and blood clotting. It occurs in milk, vegetables, and egg yolk.

Calcium Acetate The calcium salt of acetic acid which functions as a sequestrant and mold control agent. It contains approximately 25 percent calcium. It is a white odorless powder which is readily soluble in water with a solubility of approximately 37 g in 100 g water at 0°C. Its solubility decreases with increasing temperature, with a solubility of approximately 29 g in 100 g of H_2O at 100°C.

Calcium Acid Phosphate See **Monocalcium Phosphate.**

Calcium Alginate The calcium salt of alginic acid which functions as a stabilizer and thickener. The partial obtainment of calcium alginate by the reaction of the water-soluble sodium alginate with calcium ions is used to obtain viscosity and gel formation. It is used in icings, imitation pulp, dessert gels, and fabricated fruits.

Calcium Ascorbate The salt of ascorbic acid which is a white to slightly yellow crystalline powder. It is soluble in water and the pH of a 10 percent solution is 6.8 to 7.4. It functions as an antioxidant and preservative. See **Ascorbic Acid.**

Calcium Biphosphate See **Monocalcium Phosphate.**

Calcium Bromate A dough conditioner and maturing and bleaching agent which exists as a white crystalline powder. It is very soluble in water and is used in flour and dough.

Calcium Carbonate The calcium salt of carbonic acid which is used as an anticaking agent and dough strengthener. It is available in varying particle sizes ranging from coarse to fine powder. It is practically insoluble in water and alcohol, but the presence of any ammonium salt or carbon dioxide increases its solubility while the presence of any alkali hydroxide reduces its solubility. It has a pH of 9 to 9.5. It is the primary source of lime (calcium oxide) which is made by heating limestone in a furnace. Calcium carbonate is used as a filler in baking powder, for calcium enrichment, as a mild buffering agent in doughs, as a source or calcium ions in dry mix desserts, and as a neutralizer in antacids. It is also termed limestone.

Calcium Carrageenan See **Carrageenan.**

Calcium Caseinate The calcium salt of casein. Properties include low viscosity, settling out of water, opaqueness, no heat stability, and chalky texture. It contains the range of essential amino acids present in sodium caseinate but has a higher concentration of calcium. It is useful in applications requiring low absorption properties. It is used as a protein source in imitation cheese, and in special diet foods to replace sodium caseinate where sodium must be restricted. It is used to improve the whipping properties of vegetable whipped toppings, and as a binder. See **Caseinates.**

Calcium Chloride A general purpose food additive, the anhydrous form being readily soluble in water with a solubility of 59 g in 100 ml of water at 0°C. It dissolves with the liberation of heat. It also exists as calcium chloride dihydrate, being very soluble in water with a solubility of 97 g in 100 ml at 0°C. It is used as a firming agent for canned tomatoes, potatoes, and apple slices. In evaporated milk, it is used at levels not more than 0.1 percent to adjust the salt balance so as to prevent coagulation of milk during sterilization. It is used with disodium EDTA to protect the flavor in pickles and as a source of calcium ions for reaction with alginates to form gels.

Calcium Citrate The calcium salt of citric acid which functions as a sequestrant, buffer, and firming agent. It is a white, odorless powder which is slightly soluble in water. It is used as a firming agent for peppers and lima beans, and is used to improve the baking properties of flour.

Calcium Diacetate The salt of acetic acid which is used as a preservative and sequestrant.

Calcium Disodium EDTA See **Disodium Calcium EDTA.**

Calcium Gluconate A white crystalline granule or powder that functions as a firming agent, formulation aid, sequestrant, and stabilizer. At room temperature the anhydrous form has a solubility of approximately 1 g in 30 ml of water, which improves in boiling water to approximately 1 g in 5 ml of water. It also exists as calcium gluconate (monohydrate). It is used as a source of calcium ions for sodium alginate gels, and as a calcium fortifier in baked goods, puddings, and dairy product analogs. It functions as a coagulation aid in milk and instant pudding powders, and as a means of masking the bitter aftertaste of some artificial sweeteners.

Calcium Glycerophosphate A nutrient and dietary supplement which is a white odorless powder of poor water solubility. It is used in dental impression material and baking powder.

Calcium Hydrate See **Calcium Hydroxide.**

Calcium Hydroxide A general food additive made by adding water to calcium oxide (lime). It has poor water solubility with a solubility of 0.185 g in 100 g water at 0°C. The pH of a water solution at 25°C is approximately 12.4. It is used to promote dispersion of ingredients in sauces, creamed spinach, and a frozen pea/potato dish. It is used at 0.1 percent to stabilize the potassium iodide of iodized salt, and it used to be used as a neutralizer for soured cream prior to buttermaking. It is also termed hydrated lime, calcium hydrate, and slaked lime.

Calcium Iodate A source of iodine that is a white powder of slight solubility in water, but greater solubility in water containing iodides or amino acids. It is more stable than the iodide form. It is used as a dough conditioner in bread and is a source of iodine in table salts.

Calcium Lactate The calcium salt of lactic acid which is soluble in water. It has a solubility of 3.4 g per 100 g of water at 20°C and is very soluble in hot water. It is available as a monohydrate, trihydrate, and pentahydrate. The trihydrate and pentahydrate have solubilities of 9 g in 100 ml H_20 at 25°C. It contains approximately 14 percent calcium. It is used to stabilize and improve the texture of canned fruits and vegetables by converting the labile pectin to the less soluble calcium pectate. It thereby prevents structural collapse during cooking. It is used in angel food cake, whipped toppings, and meringues to increase protein extensibility which results in an increase of foam volume. It is also used in calcium fortified foods such as infant foods and is used to improve the properties of dry milk powder.

Calcium Lactobionate The calcium salt of lactobionic acid (4-(B, D-galactosido)-D-gluconic acid) produced by the oxidation of lactose. It is soluble in water and is used as a firming agent in dry pudding mixes.

Calcium Oxide A general food additive consisting of white granules or powder of poor water solubility. It is obtained by heating limestone (calcium carbonate) in a furnace. It is also termed lime or quicklime. It is used as an anticaking agent, firming agent, and nutritive supplement in applications such as grain products and soft candy.

Calcium Pantothenate A nutrient and dietary supplement which is the calcium chloride double salt of calcium pantothenate. It is a white powder of bitter taste and has a solubility of 1 g in 3 ml of water. It is used in special dietary foods.

Calcium Pectinate The salt of pectin which is obtained from citrus or apple fruit. It results from the interaction of low-methoxyl pectin with calcium ions to form a gel. It is used as a gel coating for meat products and to form food gels. See **Low-Methoxyl Pectin.**

Calcium Peroxide A dough conditioner which exists as a white or yellowish powder or granule that is insoluble in water. It improves dough strength, grain, and texture, and increases absorption and crumb resiliency. It is used in bakery products.

Calcium Phosphate A compound existing in several forms which include the monobasic, dibasic, and tribasic forms of calcium phosphate. As calcium phosphate monobasic, also termed monocalcium phosphate, calcium biphosphate, and acid calcium phosphate, it is used as a leavening agent and acidulant. Calcium phosphate dibasic, also termed dicalcium phosphate dihydrate, is used as a dough conditioner and mineral supplement. Calcium phosphate tribasic, also termed tricalcium phosphate and precipitated calcium phosphate, is used as an anticaking agent, mineral supplement, and conditioning agent.

Calcium Phosphate, Dibasic Anhydrous See **Dicalcium Phosphate, Anhydrous.**

Calcium Phosphate, Dibasic Hydrous See **Dicalcium Phosphate, Dihydrate.**

Calcium Phosphate, Monobasic See **Monocalcium Phosphate.**

Calcium Phosphate Tribasic See **Tricalcium Phosphate.**

Calcium Propionate The salt of propionic acid which functions as a preservative. It is effective against mold, has limited activity against bacteria, and no activity against yeast. It is soluble in water with a solubility of 49 g per 100 ml of water at 0°C and insoluble in alcohol. It is less soluble than sodium propionate. Its optimum effectiveness is up to pH 5.0 and it has reduced action above pH 6.0. It is used in bakery products, breads, and pizza crust to protect against mold and "rope." It is also used in cold-pack cheese food and pie fillings. Typical usage level is 0.2 to 0.3 percent and 0.1 to 0.4 percent based on flour weight.

Calcium Pyrophosphate A nutrient and dietary supplement that exists as a white odorless powder, insoluble in water. It is used in dental impression materials and as a buffer.

Calcium Saccharin A sweetener that is the calcium form of saccharin, existing as white crystals or powder with a solubility of 1 g in 1.5 ml of water. Sodium saccharin is the more common form, but calcium saccharin is available for nonsodium diets. In this form it is about 500 times as sweet as sucrose. See **Saccharin.**

Calcium Silicate An anticaking agent that exists in different forms, which are insoluble in water. It is used in salt to enhance flowability under extremely high humidity conditions. It is also used in baking powder and fabricated chips to absorb water or other liquids.

Calcium Sorbate A preservative that is the calcium salt of sorbic acid. It is not the most common form. Its solubility in water or fat is very limited and therefore it is used on surfaces for preservation. It is permitted in cheese and wrapping materials.

Calcium Stearate The calcium salt of stearic acid which functions as an anticaking agent, binder, and emulsifier. It is used in garlic salt, dry molasses, vanilla and vanilla-vanillin powder, salad dressing mix, and meat tenderizer. It can be used for mold release in the tableting of pressed candies.

Calcium Stearoyl-2-Lactylate A mixture of calcium salts of stearoyl lactylic acids and minor proportions of other calcium salts of related acids. It is manufactured by the reaction of stearic acid and lactic acid and conversion to the calcium salts, and is used as follows: as a dough conditioner in yeast-leavened bakery products and prepared mixes for yeast-leavened bakery products in an amount not to exceed 0.5 part for each 100 parts by weight of flour used; as a whipping agent in liquid and frozen egg white at a level not to exceed 0.05 percent; in whipped vegetable oil topping at a level not to exceed 0.3 percent of the weight of the finished whipped vegetable oil topping; and as a conditioning agent in dehydrated potatoes in an amount not to exceed 0.5 percent by weight.

Calcium Sulfate A general additive available as both calcium sulfate anhydrous, made by the high-temperature calcining of gypsum which is then ground and separated, and calcium sulfate dihydrate, which is made by grinding and separating gypsum containing about 20 percent water of crystallization. Calcium sulfate anhydrous contains approximately 29 percent calcium, and calcium sulfate dihydrate contains approximately 23 percent calcium. It is used, among other things, as a filler and baking powder for standardization purposes; a firming agent in canned potatoes, tomatoes, carrots, lima beans, and peppers; in dough as a source of calcium ions (because the absence of calcium ions causes bread dough to be soft and sticky and to produce bread of poor quality); in soft-serve ice cream to produce dryness and stiffness; as a calcium ion source for reaction with alginates to form dessert gels; and as a calcium source for food enrichment.

Cananga Oil A flavoring agent. It is a yellow liquid with a harsh, flowery odor. It is soluble in most fixed oils and mineral oil, and insoluble in glycerin and propylene glycol. It is obtained by distillation of flowers of *Cananga odorata* Hook and Thomas (tree of the *Anonaceae* family).

Candelilla Wax A lubricant and surface finishing agent obtained from the candelilla plant. It is a hard, yellowish-brown, opaque-to-translucent wax. Candelilla wax is prepared by immersing the plants in boiling water containing sulfuric acid and skimming off the wax that rises to the surface. It is composed of about 50 percent hydrocarbons with smaller amounts of esters and free acids. It is used in chewing gum and hard candy.

Cane Sugar See **Sugar.**

Canola Oil See **Rapeseed Oil, Low Erucic Acid**

Canthaxanthin A synthetic red colorant that is the carotenoid of most intense red color. It is available in oil-soluble, oil-dispersible, and water-dispersible forms. It has fair pH, heat, light, and chemical stability with a low tinctorial strength. Unlike the carotenoids beta-carotene and beta-apo-8′-carotenal, it does not possess vitamin A activity. Maximum usage level is 66 parts per million. Uses include carbonated soft drinks, salad dressing, and spaghetti sauce.

Caprylic Acid (Octanoic Acid) A flavoring agent considered to be a short or medium chain fatty acid. It occurs normally in various foods and is commercially prepared by oxidation of *n*-octanol or by fermentation and fractional distillation of the volatile fatty acids present. It is used in maximum levels, as served, of 0.13 percent for baked goods; 0.04 percent for cheeses; 0.005 percent for fats and oils, frozen dairy desserts, gelatins and puddings, meat products, and soft candy; 0.016 percent for snack foods; and 0.001 percent or less for all other food categories.

Caramel A colorant that is an amorphous, dark-brown product resulting from the controlled heat treatment of carbohydrates such as dextrose, sucrose, and malt syrup. It is available in liquid and powdered forms, providing shades of brown. In coloring a food with caramel, the food components must have the same charge as the particles of caramel, otherwise the particles will attract one another and precipitate out. Caramel can exist as several types, for example, acid-proof caramel of negative charge which is used in carbonated beverages; acidified solutions, bakers' and confectioners' caramel which are used in baked goods; and dried caramel for dry mixes. Major uses are in coloring beverages such as colas and root beers and in baked goods.

Caraway A spice that is a seed obtained from the tree *Carum carvi.* It has a flavor similar to dill. It is used in rolls, bread, meats, and some cheeses.

Carbon Dioxide A gas obtained during fermentation of glucose (grain sugar) to ethyl alcohol. It is used in pressure-packed foods as a propellant or aerating agent and is also used in the carbonation of beverages. It is released as a result of the acid carbonate reaction of leavening agents in baked goods to produce an increase in volume. As a solid, it is termed dry ice and is used for freezing and chilling.

Carbonated Water A beverage made by absorbing carbon dioxide in water. The carbon dioxide influences flavor because increased carbonation increases mouth feel. Gas retention is more common in low-calorie-type beverages because of the absence of sugar solids.

Carboxymethylcellulose (CMC) A gum that is a water-soluble cellulose ether manufactured by reacting sodium monochloroacetate with alkali cellulose to form sodium carboxymethylcellulose. It dissolves in hot or cold water and is fairly stable over a pH range of 5.0 to 10.0, but acidification below pH 5.0 will reduce the viscosity and stability except in a special acid-stable type of CMC. A variety of types are available which differ in viscosity and degree of substitution (the number of sodium groups per unit). It functions as a thickener, stabilizer, binder, film former, and suspending agent. It is used in a variety of foods to include dressings, ice cream, baked goods, puddings, and sauces. The usage range is from 0.05 to 0.5 percent.

Cardamon A spice that is a dried, ripe seed of *Elettaria cardamomum,* a biennial plant. It has a pungent aroma and is reddish-brown in color. The flavor is sweet and spicy with a camphoraceous note. It is used in whole form to flavor hot fruit punches, pickles, and marinades. It is used in the ground form in bread, cookies, desserts, and meats.

Carmine The red colorant aluminum lake of carminic acid which is the coloring pigment obtained from dried bodies of the female insect *Coccus* *cacti.* It is brilliant red to purplish in color, having a low tinctorial strength, and can be solubilized in ammonia. It is used in a pink color in coatings.

Carnauba Wax A general purpose food additive that is a hard and brittle wax. It is obtained from leaf buds and leaves of the Brazilian wax palm *Copernicia cerifera.* It is the hardest wax known and is used in candy glaze.

Carob A cocoa substitute obtained from the pods of the carob tree *Ceratonia Siliqua.* The pods are kibbled, roasted, and ground into a powder which is similar in appearance and fragrance to cocoa powder. Carob powder has less than 1 percent fat and 42 to 48 percent sugar, while cocoa powder has approximately 23 percent fat and 5 percent sugar. Cocoa does not contain any measurable amounts of fructose so the presence of carob in cocoa can be detected by the presence of fructose. It is used in candy, drinks, bakery products, and dairy applications, and as a single ingredient in health food products.

Carob Gum See **Locust Bean Gum.**

Carotene A colorant and provitamin, being a hydrocarbon which is one of two subgroups of the carotenoids (yellow, orange, or red pigments). The other subgroup is xanthophylls. Carotene functions as a colorant with beta-apo-8′-carotenal being a red-orange carotenoid and beta-carotene being a yellow carotenoid. It is also a vitamin A precursor that is converted by the body to vitamin A. It is used in ice cream, cheese, and other dairy products.

Carrageenan A gum that is a seaweed extract obtained from red seaweed *Chondrus crispus* or *Gigartina mammillosa.* It exists as various salts or mixed salts of a sulfate ester. It is classified mainly as kappa, iota, and lambda types which differ in solubility and gelling properties. The kappa and iota types require hot water (above 71°C) for complete solubility and can form thermally reversible gels in the presence of potassium and calcium cations, respectively. The kappa gels are brittle with syneresis while the iota gels are more elastic without syneresis. The lambda type is cold-water soluble and does not form gels. Kappa and iota carrageenan are very reactive with milk protein products. Carrageenan is used to stabilize milk protein at 0.01 to 0.05 percent and to form water gels at 0.5 to 1.0 percent. Its uses include dairy products, water gel desserts, and low-calorie jellies. A typical use level in water systems is 0.2 to 1.0 percent and milk systems is 0.01 to 0.25 percent.

Carubin See **Locust Bean Gum.**

Carvacrol A flavoring agent that is a colorless to pale yellow liquid. It has a spicy and pungent odor, resembling thymol. It is insoluble in water and soluble in alcohol and ether. It is a mixture of the isomeric carvacrols (isopropyl o-creols), and is obtained by chemical synthesis. It is also an ingredient of savory, a fragrant herb in nature.

Casein The principal milk protein which is prepared commercially from skim milk by the precipitation with lactic, hydrochloric, or sulfuric acid. It can also be produced by the use of lactic-acid-producing bacteria. Caseins are usually identified according to the acid used, such a lactic acid casein, hydrochloric acid casein, and sulfuric acid casein. The principal form in which casein is used is casein salts, of which sodium and calcium caseinate are the most common. Rennet casein is obtained from skim milk by the precipitation with a rennet-type enzyme. Casein is used in the protein fortification of cereals and bread, and in fabricated cheeses.

Caseinates Salts of casein that are produced by neutralizing acid casein to pH 6.7 with calcium or sodium hydroxide, producing the most common forms, which are calcium caseinate or sodium caseinate. Other forms of casein are potassium and ammonium caseinate. The caseinates provide a source of protein and function as emulsifiers, water binders, and whipping aids. The relative water absorption of casein salts is: calcium caseinate—130 percent, potassium caseinate—200 percent, sodium caseinate—250 percent. Its uses include processed meats, whipped toppings, coffee whiteners, egg substitutes, and diet foods.

Castor Oil A release and antisticking agent used in hard candy production. Its concentration is not to exceed 500 parts per million (ppm). It is used in vitamin and mineral tablets, and as a component of protective coatings.

Cayenne Pepper See **Pepper, Cayenne.**

Celery Seed A spice made from the dried, ripe fruit of the herb *Apium graveolens,* related to the parsley family. It is used lightly so as not to dominate in flavor. It is used in sauces, salads, meats, and soups.

Cellulose A carbohydrate polymer made up of glucose units. It consists of fibrous particles and is used as a fiber source and bulking agent in low-calorie formulations.

Cheese Culture Bacteria used in the coagulation of the milk protein casein in the formation of cheese. It converts milk into cheese curd by the reduction of pH followed by processing to precipitate the protein as a curd.

Cheese Powder A dry form of cheese prepared by slurrying cheese in water to 35 to 45 percent solids and further processing into a powder form. Cheese powders are water soluble. They are used in instant soups, dry spaghetti sauce, dry sauces, and snack foods.

Chelating Agents See **Sequestrants.**

Chervil A spice derived from the plant *Anthriscus cerefolium* which is related to the parsley family. It is used in soufflés, sauces, meats, and fish.

Chewing Gum Base A base, containing masticatory substances such as chicle, used in the manufacture of chewing gum.

Chicle A natural masticatory substance of vegetable origin which is used in chewing gum base. It is the latex of the sapodilla tree, obtained by cutting the bark to yield the latex which is boiled to remove about two-thirds of the water. The resulting semisolid mass is molded into chicle blocks which form the base for chewing gum.

Chilte A substance of vegetable origin used as a masticatory substance in chewing gum base.

Chiquibul A substance of vegetable origin used as a masticatory substance in chewing gum base.

Chives A spice from the *Allium schoenoprasum* plant whose slender rush-like green leaves are chopped and used to provide a subtle onion flavor and to enhance food appearance. It is also used as a garnish and topping.

Chlorine A gas used to age and bleach flour.

Chlorine Dioxide A gas used in bleaching and aging flour. It acts on the flour almost instantly, resulting in improved color and dough properties. Because usage levels are low the bleaching action is limited.

Chloropentafluoroethane A propellant and aerating agent for foamed or sprayed foods.

Chlorophyll A colorant that is a green pigment present in all green plants. It is used in sausage casings, oleomargarine, and shortening.

Chocolate A solid or semiplastic food made from chocolate liquor derived from cocoa nibs, which are obtained from the cocoa bean. Chocolate contains more fat and less protein than cocoa. The products derived from chocolate include bitter or plain chocolate; sweet chocolate containing sugar, milk, flavoring, and cocoa butter; and milk chocolate, which is made from sweet or bitter chocolate plus a milk source with or without cocoa butter and flavoring. It is used as a flavor in candy, dairy products, and baked goods.

Chocolate Liquor See **Cocoa Liquor**

Cholic Acid An emulsifier that exists as colorless plates or a white crystalline powder which has a bitter taste with a sweetish aftertaste. It is slightly soluble in water. It functions as an emulsifying agent in egg white.

Choline A substance grouped as a member of the vitamin B complex, although not a vitamin by definition. It is water soluble and is important in nerve function and fat metabolism. It occurs in egg yolk, beef liver, and grains.

Cider Vinegar The product made by the alcoholic and subsequent acetous fermentation of apple juice or concentrate thereof. It contains not less than 4 g acetic acid in 100 cm^3 at 20°C. It has a light to medium amber color. It is used in salad dressings, mayonnaise, and sauces. The term *vinegar* refers to cider vinegar, also termed apple vinegar.

Cinnamic Acid (3-Phenylpropenoic Acid) A flavoring agent that consists of crystalline scales, white in color, with an odor resembling honey and flowers. It is slightly soluble in water, soluble in alcohol, chloroform, acetic acid, acetone, benzene, and most oils, and alkali salts soluble in water. It is obtained by chemical synthesis.

Cinnamon A spice made from the dried bark of the evergreen tree *Cinnamomum cassia*. Commercial types are Saigon Cassia and Batavia Cassia. Ceylon cinnamon is the dried inner bark of shoots of *C. zeylanicum* Nees. In the ground form it is used in beverages, desserts, and fruits while in the stick form it is used in beverages, meats, and fruits.

Cinnamyl Anthranilate A flavoring agent that is a powder which may be red or yellow. It has an odor resembling anthranilates, fruity and characteristically balsamic. It is insoluble in water, and soluble in alcohol, chloroform, and ether. It is obtained by chemical synthesis.

Cinnamyl Isobutyrate A synthetic flavoring agent that is a moderately stable, colorless to light yellow liquid of dry fruity color. It is stored in glass or tin. It is used to give a lift to jasmine with applications in baked goods and candy at 8 pars per million.

Citral A liquid flavoring agent, light yellow in color with a citrus odor. It occurs in lemon and lemongrass oils. It is usually obtained from citral-containing oils by chemical means but may also be prepared synthetically. It is soluble in fixed oils, mineral oil, and propylene glycol. It is moderately stable and should be stored in glass, tin, or resin-lined containers. It is used in flavors for lemon with applications in candy, baked goods, and ice cream at 20 to 40 parts per million. It is also termed 2,6-dimethyl-octadian-2-6-al-8.

Citric Acid An acidulant and antioxidant produced by mold fermentation of sugar solutions and by extraction from lemon juice, lime juice, and pineapple canning residue. It is the predominant acid in oranges, lemons, and limes. It exists in anhydrous and monohydrate forms. The anhydrous form is crystallized in hot solutions and the monohydrate form is crystallized from cold (below 36.5°C) solutions. Anhydrous citric acid has a solubility of 146 g and monohydrate citric acid has a solubility of 175 g per 100 ml of distilled water at 20°C. A 1 percent solution has a pH of 2.3 at 25°C. It is a hygroscopic, strong acid of tart flavor. It is used as an acidulant in fruit drinks and carbonated beverages at 0.25 to 0.40 percent, in cheese at 3 to 4 percent, and in jellies. It is used as an antioxidant in instant potatoes, wheat chips, and potato sticks, where it prevents spoilage by trapping the metal ions. It is used in combination with antioxidants in the processing of fresh frozen fruits to prevent discoloration.

Citronellal (3,7-Dimethyl-6-Octen-1-Al) A flavoring agent that is a liquid, faintly-yellow with an intense odor resembling lemon, citronella, and rose. It is soluble in alcohol and most fixed oils, slightly soluble in mineral oil and propylene glycol, and insoluble in water and glycerin. It is obtained by chemical synthesis; the aldehyde may be obtained from natural oils, such as citronella oil.

Citronellyl Propionate A synthetic flavoring agent that is a moderately stable, colorless liquid of light rose-fruity odor. It is practically insoluble in water but is miscible with alcohol. It is stored in glass or tin containers.

It has application in baked goods, candy, beverages, and ice cream at 3 to 19 pats per million.

Citrus Oil A flavorant obtained by pressing the oil from the rind of citrus fruits. It is largely composed of terpenes and sesquiterpenes plus the flavor-imparting oxygenated components. It is partly water soluble, not stable, and is used in beverages.

Clarified Butter See **Butter, Clarified.**

Clarifying Agents See **Processing Aids.**

Clear Flour The portion of straight flour (all the flour that can be milled from a wheat blend) that remains after the removal of the patent streams. Clear flours from hard wheat are generally high in ash, dark in color, and high in protein. It is used to increase the strength of flour and is used in rye, dark bread, and pastries.

Clouding Agents See **Processing Aids.**

Clove A spice that is the unripened bud from the clove tree *Eugenia caryophyllata* Thumb. It is very pungent and is used in the whole form in fruit punches, relishes, marinades, and sauces. In the ground form, it is used in cakes, cookies, and meat sauces.

Coarse-Ground Wheat See **Crushed Wheat.**

Cochineal A red colorant extracted from the dried bodies of the female insect *Coccus cacti.* The coloring is carminic acid in which the water-soluble extract is cochineal. It precipitates at pH 3, has good stability at pH 4, and excellent stability at pH 5 to 8. It has low tinctorial strength and has excellent stability to heat and light. It is also stable in retorted protein systems where other food dyes are unstable. It is used in foods requiring red coloring.

Cocoa Butter The fat obtained by pressing chocolate liquor, obtained from roasted cocoa nibs, to yield cocoa butter and presscake. It has a melting point of approximately 33°C but is a hard, brittle solid at room temperature. It is used in the manufacture of coatings for candies, the coatings consisting mainly of mixtures of roasted cocoa nibs, sugar, and cocoa butter. It is also used in confections.

Cocoa Liquor The liquor obtained by the grinding of cocoa nibs from the cocoa bean. The liquor converts into cocoa powder and cocoa butter as end products. It is a primary ingredient in chocolate manufacture. It is also termed chocolate liquor.

Cocoa Powder The powder produced by the grinding, pulverizing, and air classification of the cocoa presscake, which is obtained by compressing the cocoa liquor, obtained from cocoa nibs, into a presscake and cocoa butter. There are two main types of powder—alkalized and natural. The alkalized (Dutch processed) has a pH range of 6.5 to 8.1, a red-brown shade which tends to develop red-brown end products, and a mild flavor. It is used in beverages, retail cocoa powder, puddings, and ice cream. The natural has a pH range of 5.2 to 5.9 and a yellow-orange color with a tendency to produce light brown end products. It is used in the baking industry to impart color and flavor and also used in candy, syrups, and toppings.

Coconut The nut obtained from the coconut palm. It provides a source of coconut meat and coconut oil.

Coconut, Desiccated The dried coconut meat whose reduced moisture content increases its stability. It is available in various shapes and sizes. It is used to impart flavor in desserts, baked goods, and candies.

Coconut Oil The oil obtained from the kernel of the nuts of the coconut palm. It has a sharp melting character (narrow plastic range) in that it changes abruptly from a hard, brittle solid to a clear oil with a temperature change of a few degrees, and the transition occurs at room temperature range. It melts at 25°C and is more completely solid than butter at 10°C. These properties make it suited for the preparation of shortenings where brittleness and a large change in consistency with a small temperature change are undesirable. Partially hydrogenated coconut oil has hydrogen added to part of the unsaturated carbon bonds to provide a more solid consistency. It is used in confections, baked goods, and margarine.

Collagen A protein that is the principal constituent of connective tissue and bones of vertebrates; it can be converted to gelatin and glue by boiling in water.

Colors and Coloring Adjuncts Substances used to impart, preserve, or enhance the color or shading of a food, including color stabilizers, color fixatives, color-retention agents, etc. Legally, they are usually designated artificial (synthetic) or natural, which indicates that they are, respectively, synthetically manufactured or obtained from natural sources. Synthetic color additives "certified" by the Food and Drug Administration are designated FD&C (Food, Drug, and Cosmetic). Those acceptable food colors not designated "certified" are designated "approved" and consist of natural organic and synthetic inorganic colorants used in certain applications.

Confectionary Fat A fat that is hard at room temperature and soft at body temperature, such as hydrogenated coconut oil or cocoa butter.

Copper A metal necessary for the maintenance of normal erythropoiesis and the prevention of iron deficiency anemia, iron being essential in hemoglobin synthesis.

Copper Gluconate A light blue powder used as a dietary supplement.

Copper Sulfate (Cupric Sulfate) A nutrient supplement and processing aid most often used in the pentahydrate form. This form occurs as large, deep blue or ultramarine, triclinic crystals, as blue granules, or as a light blue powder. The ingredient is prepared by the reaction of sulfuric acid with cupric oxide or with copper metal. Copper sulfate may be used in infant formula.

Coriander A spice that is the dried, ripe fruit of *Coriandrum sativum* L. It has a pleasing, aromatic taste. It is used in sausage, variety meats, and curry powder in the ground form, and in pickles, baked goods, and stuffing in the whole form.

Corn The maize grain, which is the source of various ingredients. It is used in the kernel form for food; it is dry milled into flour, grits, and meal, and it is wet milled into starches, dextrins, dextrose, and other by-products. The kernel consists of four basic parts which are the starch section, corn germ, gluten, and hull. The starch section comprises approximately 61 percent of the kernel, while the corn germ comprises approximately 4 percent of the kernel. The term *corn* refers to other cereal crops in different areas of the world.

Corn Bran A dry-milled product of high fiber content obtained from corn. It can be used to increase the fiber content of breads, cookies, and cereals, and to thicken gravies and soups.

Corn Flour A finely ground flour made from milling and shifting maize or obtained as a by-product of cornmeal. It is used as pancake flour.

Corn Gluten (Corn Gluten Meal) A nutrient supplement which is the principal protein component of corn endosperm. It consists mainly of zein and glutelin. Corn gluten is a byproduct of the wet milling of corn for starch. The gluten fraction is washed to remove residual water soluble proteins. Corn gluten is also produced as a byproduct during the conversion of the starch in whole or various fractions of dry milled corn syrups. The ingredient is used in food with no limitation other than current good manufacturing practice.

Cornmeal A ground corn of specified mesh profile that is made from white or yellow maize. It is used in cornbread mix.

Corn Oil The oil obtained from the germ of the maize plant. The unsaturated fatty acids linoleic and oleic make up 80 to 85 percent of the total fatty acids. The tocopherols prevent the oil from oxidizing rapidly. It has a low melting point of −18 to −10°C. It is used in mayonnaise, margarine, salad oil, and bakery products.

Corn Silk and Corn Silk Extract Flavor agents used in baked goods and baking mixes (30 ppm), nonalcoholic beverages (20 ppm), frozen dairy desserts (10 ppm), soft candy (20 ppm), and all other food catagories (4 ppm). Corn silk is the fresh styles and stigmas of *Zea mays* L. collected when the corn is in milk. The filaments are extracted with dilute ethanol to produce corn silk extract. The extract may be concentrated at a temperature not exceeding 60°C.

Cornstarch The starch made from the endosperm of corn, containing amylose and amylopectin starch molecules. When starch is heated in water it forms a viscous, opaque paste. The paste forms semisolid gels upon cooling and has the ability to form strong adhesive films when spread and dried. Cornstarch is not freeze-thaw stable and is used widely except when clarity or the lack of gel formation is desired. It exists as fine or coarse powders. The coarse starch is sometimes termed pearl starch. It is used in sauces, gravies, puddings, pie fillings, and salad dressings. The typical usage level is 1 to 5 percent. It is also termed common, regular, or unmodified cornstarch.

Cornstarch, Acid-Modified A starch produced by treating suspended cornstarch in water with dilute mineral acid at high temperatures for varying time periods. This is followed by neutralization with sodium carbonate upon obtainment of the desired viscosity. This produces starches that have decreased viscosity when warm but still form gels when cooled. It is used in the manufacture of starch-based gum candies.

It is also termed thin-boiling starch. Esters and ethers can be formed in which only one end of the addition molecule is attached to the starch molecule. These starches have freeze-thaw stability, shear resistance, and acid resistance, and are used in sauces, gravies, and frozen foods.

Cornstarch, Oxidized See **Oxidized Cornstarch.**

Corn Sugar See **Dextrose.**

Corn Sugar Vinegar The product made by the alcoholic and subsequent acetous fermentation of corn sugars according to federal regulations. It is of amber color and has a minimum of 4 percent acid (expressed as acetic acid). It functions as an acidulant in foods.

Corn Syrup A corn sweetener that is a viscous liquid containing maltose, dextrin, dextrose, and other polysaccharides. It is obtained from the incomplete hydrolysis of cornstarch. It is classified according to the degree of conversion which is expressed as the dextrose equivalent (DE), which is the measure of sweetness of the corn syrup as compared to that of a sucrose syrup. Generally the greater the degree of conversion, the sweeter the syrup. Corn syrup is used as a replacement for sucrose but is less sweet than sucrose. It can control crystallization in candy making, contribute body in ice cream, and provide pliability in confections. It is also termed glucose syrup.

Corn Syrup Solids The dry form of corn syrup used where it is impractical to use the liquid syrup. See **Corn Syrup.**

Cracked Wheat The wheat prepared by cracking or cutting cleaned wheat, other than durum wheat and red durum wheat, into angular fragments. The proportions of the natural constituents, other than moisture, remain unaltered. The moisture content does not exceed 15 percent.

Cracker Flour Flour that is long patented or straight grades of soft wheat flour, containing 9 to 10.5 percent protein. It is of low absorption and has short mixing requirements.

Cranberry Extract A natural red colorant with good pH stability and fair heat, light, and chemical stability. The anthocyanidin pigments in cranberry are peonidin and cyanidin. The extract has low tinctorial strength and good stability at pH 3 to 4. Because the color is affected by pH, it can only be used in acidic mediums such as beverages.

Cream That portion of milk that is high in milkfat and will rise to the top of undisturbed milk. It is obtained by the separation of the fat fraction of the milk to concentrations ranging from 18 to 40 percent fat. Cream is labeled according to the fat content: heavy whipping cream has a minimum of 36 percent fat; light whipping cream has 30 to 36 percent fat; and light, coffee, or table cream has 18 to 30 percent fat. The lower fat creams are usually prepared by blending a high-fat cream with milk. Cream is used in ice cream mix, whipped toppings, and sauces.

Cream of Tartar The acid potassium salt of tartaric acid occurring as crystals or powder. It is relatively poorly soluble having a solubility in 100 ml of water of 0.8g at 25°C and 6.1 g at 100°C. A 1 percent solution at 30°C has a pH of 3.4. Chemical names are potassium acid tartrate, potassium hydrogen tartrate, and potassium bitartrate. It functions to complex with heavy metal ions and to regulate pH; it can have a gentle laxative action if given at adequate levels. The acidulant is used in chemical leavening to release carbon dioxide which produces the loaf volume. It has limited reactivity in the cold so when used in reduced-temperature batters it has little gas evolution during the initial mixing. At room temperatures, it has a relatively fast reaction rate. It functions as a taste regulator in sugar icing and in the controlled crystallization of toffees and fondants by the regulated inversion of sucrose. It is used in baked goods, crackers, candy, and puddings.

Cresyl Acetate (p-Tolyl Acetate) A flavoring agent that is a clear and colorless liquid with a strong, flowery odor. It is soluble in most fixed oils and propylene glycol, moderately soluble in mineral oil, and insoluble in glycerin. It is obtained by chemical synthesis.

Crown Gum A product of vegetable origin used as a masticatory substance in chewing gum base.

Crushed Wheat The wheat prepared by crushing cleaned wheat, other than durum wheat and red durum wheat. The proportions of the natural constituents, other than moisture, remain unaltered. The moisture content does not exceed 15 percent. It is also termed coarse-ground wheat.

Cumin A spice that is the dried, ripe fruit of *Cuminum cyminum* L. It is usually obtained in the ground form. It has a warm, pleasant, balsamic flavor. It is used in cheese, soups, relishes, and meats.

Cuminic Aldehyde (p-Cuminic Aldehyde; Cumaldehyde; Cuminal) A flavoring agent that is a liquid, colorless to yellow in appearance, with a strong, pungent odor resembling cumin oil. It is insoluble in water and soluble in alcohol and ether. It is obtained from cumin oil.

Curry Powder A blend of spices used as seasoning in curries, sauces, and meats. Typical spices in the blend include coriander, ginger, nutmeg, clove, cinnamon, red pepper, and onion salt.

Cyanocobalamin Vitamin B_{12}, a water-soluble vitamin required for the normal development of red blood cells. Its deficiency causes pernicious anemia. It is stable in neutral conditions and is more stable for storage than for processing conditions. It is found in meat, fish, and milk.

Cyclohexyl Acetate A synthetic flavoring agent that is a stable, colorless liquid of fruity odor. It is stored in glass, tin, or resin-lined containers. It is used for flavors such as apple, banana, blackberry, and raspberry with applications in beverages, ice cream, candy, and baked goods at 15 to 110 parts per million.

Cyclohexyl Butyrate A synthetic flavoring agent that is a stable, colorless liquid of fruity odor. It should be stored in glass, tin, or resin-lined containers. It is used in pineapple, apricot, banana, and berry flavor with applications in beverages, ice cream, and candy at 4 to 9 parts per million.

Cyclohexyl Cinnamate A synthetic flavoring agent that is a stable, colorless to light yellow liquid of fruity odor. It is stored in glass or tin containers. It is used in peach and cherry flavors with applications in ice cream, candy, and baked goods at 5 to 20 parts per million.

Cyclohexyl Formate A synthetic flavoring agent that is a moderately stable, colorless liquid of fruity odor. It should be stored in glass or tin containers. It is used in apple and plum flavors and gives a lift to fruity flavors. It has application in beverages, candy, ice cream, and baked goods at 3 to 11 parts per million.

Cyclohexyl Propionate A synthetic flavoring agent that is a stable, colorless liquid of fruity odor. It is stored in glass or tin containers. It is used in fruit flavors such as pineapple with applications in beverages, candy, ice cream, and baked goods at approximately 3 parts per million.

Cydonia Seed See **Quince Seed.**

Cysteine A nonessential amino acid that functions as a nutrient and dietary supplement. It is used in foods to prevent oxygen from destroying vitamin C and is used in doughs to reduce mixing time.

Cystine A nonessential amino acid that acts as a nutrient and dietary supplement. It is very slightly soluble in water and in alcohol. It improves the biological quality of the total protein in foods containing naturally occurring intact protein.

D

d-Limestone (d-p-Mentha-1,8,Diene; Cinene) A flavoring agent that is a liquid, colorless with a pleasant odor resembling mild citrus. It is miscible in alcohol, most fixed oils, and mineral oil; soluble in glycerin; and insoluble in water and propylene glycol. It is obtained from citrus oil.

Danish Agar See **Furcelleran**.

1-Decanol, Natural (Decyl-Alcohol) A flavoring agent that is a liquid, colorless, with a flowery odor similar to orange blossoms. It is insoluble in water and glycerin, and soluble in alcohol, ether, and mineral oil.

Decyclic Alcohol See **Decanol**.

Defoaming agents. See **Surface-Active Agents**.

Dehydroacetic Acid (DHA) A preservative that is a crystalline powder with a solubility of less than 0.1 g in 100 g H_2O at 25°C. It can undergo a variety of chemical reactions which give it utility in many applications. It is used at 0.01 to 0.5 percent for microbiological growth inhibition in various foods. It is used for cut or peeled squash, with no more than 65 parts per million remaining in or on the prepared squash.

Dextrin A carrier formed from the treatment of starch by dry heat, acid, or enzymes. It can be formed from amylose and amylopectin-type starches. The cold-water soluble dextrins are used as carriers for active ingredients such as flavors. Industrially, it is used as an adhesive. It is used as a flavor carrier in dry-mix beverages, soups, and gravy.

Dextrose A corn sweetener that is commercially made from starch by the action of heat and acids or enzymes, resulting in the complete hydrolysis of the cornstarch. There are two types of refined dextrose commercially available. Dextrose hydrate, which contains 9 percent by weight water of crystallization and is the most often used, and anhydrous dextrose, which contains less than 0.5 percent water. Dextrose is a reducing sugar and produces a high-temperature browning effect in baked goods. It is used in ice cream, bakery products, and confections. It is also termed glucose and corn sugar.

Diacetyl Tartaric Acid Esters of Mono- and Diglycerides A hydrophilic emulsifier used in oil-in-water emulsions. The connecting of glycerol with tartaric acid prior to esterification of the other part of the glycerol increases the hydrophilicity of the emulsifier. It functions as a dough conditioner in freestanding breads and rolls to strengthen the gluten which improves crumb softness, crust, and increased volume. It is used in coffee whiteners for dispersion. It is used in chocolate couverture to adjust the consistency, viscosity, and adhesion ability. In reduced-calorie breads, it reduces the quantity of shortening required and maintains volume. It is also termed acetylated tartaric acid monoglyceride, and acetyl tartrate mono- and diglyceride.

Dibasic Calcium Phosphate, Anhydrous See **Dicalcium Phosphate, Anhydrous**.

Dibasic Calcium Phosphate, Dihydrate See **Dicalcium Phosphate, Dihydrate**.

Dicalcium Phosphate, Anhydrous A mineral supplement and dough conditioner. It contains approximately 29 percent calcium. It is practically insoluble in water, with a solubility of 0.02 g per 100 ml H_2O at 25°C. It is also termed calcium phosphate, dibasic anhydrous. It is used as a mineral supplement in prepared breakfast cereals, enriched flour, and noodle products.

Dicalcium Phosphate, Dihydrate A source of calcium and phosphorus that also functions as a dough conditioner and bleaching agent. It functions as a dough conditioner in bakery products, as a bleaching agent in flour, as a source of calcium and phosphorus in cereal products, and as a source of calcium for alginate gels. It contains approximately 23 percent calcium.

It is practically insoluble in water. It is also termed dibasic calcium phosphate, dihydrate and calcium phosphate dibasic, hydrous. It is used in dessert gels, baked goods, cereals, and breakfast cereals.

Diacetyl A flavoring agent that is a clear yellow to yellowish green liquid with a strong pungent odor. It is also known as 2,3-butane-dione and is chemically synthesized from methyl ethyl ketone. It is miscible in water, glycerin, alcohol, and ether, and in very dilute water solution it has a typical buttery odor and flavor.

Diethyl Sebacate (Ethyl Sebacate) A flavoring agent that is a liquid, colorless, to pale yellow in appearance with a slight odor. It is insoluble in water and miscible in alcohol, ether, and other organic solvents. It is obtained by chemical synthesis.

Diglyceride A lipophilic emulsifier prepared by direct esterification of two fatty acids with glycerol, or by interesterification between glycerol and other triglycerides. It often occurs as a blend with monoglycerides. It is widely used in numerous foods such as ice cream, puddings, margarine, doughs, shortenings, peanut butter, and coffee whiteners. It has numerous functions including the provision of dough conditioning, prevention of fat separation, and the provision of emulsion stability and dispersibility.

Dilauryl Thiodipropionate (DLTDP) An antioxidant that exists as white crystalline flakes of sweetish ester-like odor. It is insoluble in water but soluble in inorganic solvents. It is used in fats and oils to prevent rancidity. It is used in peanut oil at a maximum usage level of 200 mg/kg.

Dill and Its Derivatives A flavoring agent that is the herb and seeds from *Anethum graveolens* L., dill (Indian), and the herb and seeds from *Anethum sowa*, D.C. Its derivatives include essential oils, oleoresins, and natural extractives obtained from these sources of dill.

Dill Seed A spice that is the dried, ripe fruit of the plant *Anethum graveolens* L. It is extremely pungent and slightly dominant. It is used in dips, spreads, sauces, and meats.

Dill Weed A spice made from the leaf of the dill plant. While dill seed has a camphorous, slightly bitter taste and fragrance, the weed has a delicate bouquet which enhances rather than dominates. It is used in meats and sauces.

2,6-Dimethyl-Octadian-2-6-al-8 See **Citral**.

Dimethylpolysiloxane An antifoaming agent used in fats and oils. It prevents foaming and spattering when oils are heated and prevents foam formation during the manufacture of wine, refined sugar, gelatin, and chewing gum. It is also termed methylpolysilicone and methyl silicone.

Dioctyl Sodium Sulfosuccinate A wetting and emulsifying agent that is slowly soluble in water, having a solubility of 1 g in 70 ml of water. It functions as a wetting agent in fumaric acid-containing powdered fruit drinks to help the acid dissolve in water. It is used as a stabilizing agent on gums at not more than 0.5 percent by weight of the gum. It is used as a flavor potentiator in canned milk where it improves and maintains the flavor of the sterilized milk during storage. It also functions as a processing aid in the manufacture of unrefined sugar.

Dipotassium Monohydrogen Orthophosphate See **Dipotassium Phosphate**.

Dipotassium Monophosphate See **Dipotassium Phosphate**.

Dipotassium Phosphate The dipotassium salt of phosphoric acid which functions as a stabilizing salt, buffer, and sequestrant. It is mildly alkaline with a pH of 9 and is soluble in water with a solubility of 170 g per 100 ml of water at 25°C. It improves the colloidal solubility of proteins. It acts as a buffer against variation in pH. For example, it is used in coffee whiteners as a buffer against pH variation in hot coffee and to prevent feathering. It also functions as an emulsifier in specified cheeses and as a buffering agent for processed foods. It is also termed dipotassium monohydrogen orthophosphate, potassium phosphate dibasic, and dipotassium monophosphate.

Disodium Calcium EDTA A sequestrant and chelating agent whose complete name is disodium calcium ethylenediamine tetraacetate. It is a nonhygroscopic powder that is colorless, odorless, and tasteless at recommended use levels. A 1 percent solution has a pH of 6.5 to 7.5. It is used to control the reaction of trace metals with some organic and inorganic components in food; to prevent deterioration of color, texture, and development of precipitates; and to prevent oxidation. Its function is comparable to that of disodium dihydrogen EDTA. See **EDTA**.

Disodium Dihydrogen EDTA A sequestrant and chelating agent whose complete name is disodium ethylenediamine tetraacetate. It is a nonhygroscopic powder that is colorless, odorless, and tasteless at recommended use levels. A 1 percent solution has a pH of 4.3 to 4.7. It is used to control the reaction of trace metals to include calcium and magnesium with other organic and inorganic components in food to prevent deterioration of color, texture, and development of precipitates and to prevent oxidation. Its function is comparable to that of disodium calcium EDTA. See **EDTA**.

Disodium Dihydrogen Pyrophosphate See **Sodium Acid Pyrophosphate**.

Disodium Diphosphate See **Sodium Acid Pyrophosphate**.

Disodium Guanylate (Sodium 5′-Guanylate; Disodium Guanosine-5′-Monophosphate) A flavor enhancer which is a crystalline powder, colorless or white, and has characteristic taste. It is soluble in water, sparingly soluble in alcohol, and practically insoluble in ether. It is obtained by chemical synthesis.

Disodium 5′-Inosinate A flavor enhancer which performs as disodium guanylate does, but only when present at approximately twice the level. See **Disodium 5′-Guanylate**.

Disodium Monohydrogen Orthophosphate See **Disodium Phosphate**.

Disodium Monohydrogen Orthophosphate Dihydrate See **Disodium Phosphate**.

Disodium Monophosphate See **Disodium Phosphate**.

Disodium Phosphate The disodium salt of phosphoric acid which functions as a protein stabilizer, buffer, dispersant, and coagulation accelerator. It is mildly alkaline with a 1 percent solution having a pH of 9.2. It is moderately soluble in water with a solubility of 12 g in 100 ml at 25°C. It is used in farina and macaroni to shorten the cooking time by making the particles swell faster and cook more thoroughly. In evaporated milk it acts as a buffer and prevents gelation, also acting as a buffer in coffee whiteners. It is an accelerator of the setting time in instant puddings. In cream sauce and whipped products it functions as a dispersant by producing a swelling of protein. It is also termed disodium monohydrogen orthophosphate; sodium phosphate, dibasic; and disodium monophosphate.

Disodium Phosphate, Duohydrate An emulsifier, buffer, and mineral supplement. It is moderately soluble in water with a solubility of 15 g in

100 ml of water at 25°C. A 1 percent solution has a pH of 9.1. It is used in processed cheese for uniform texture and smoothness. It is also termed disodium phosphate, dihydrate and sodium phosphate dibasic, dihydrate.

Disodium Tartrate See **Sodium Tartrate**.

Dispersants See **Surface-Active Agents**.

Distilled Monoglyceride An emulsifier containing a minimum of 90 percent monoglyceride derived from edible fat and glycerine. It is an active monoglyceride produced by distillation to obtain the monoglyceride fraction, which is the part that functions as an emulsifier or food quality improver. Commercially termed monoglycerides also contain diglycerides, triglycerides, and so on. It is used in margarine, peanut butter, shortenings, bakery goods, and whipped desserts to improve texture and consistency. Typical usage levels are 0.1 to 1.0 percent.

Distilled Vinegar See **Vinegar, Distilled**.

2,6-Di-Tert-Butyl-Para-Cresol See **Butylated Hydroxytoluene**.

Dodecyl Gallate An antioxidant used in cream cheese, instant mashed potatoes, margarine, fats, and oils.

Dough Conditioner A blend of minerals used in baked goods. It is usually contained within yeast foods as a blend of calcium salts, sulfates, and phosphates which toughen the gluten. Usage of hard water generally results in better breads so the minerals serve to minimize the effect of variables in water conditions. It is also termed yeast food.

Dough Strengtheners Substances used to modify starch and gluten, thereby producing a more stable dough.

Dried Buttermilk See **Buttermilk, Dried**.

Dried Milk See **Whole Milk Solids**.

Dried Skim Milk See **Milk Solids—Not-Fat**.

Dry Ice See **Carbon Dioxide**.

Dry Whole Milk See **Milk Powder**.

Durum Flour The fine powder obtained from durum wheat, which is fine enough to pass through a number 100 U.S. sieve. It is used principally in macaroni and spaghetti products because it provides the desired texture and consistency. See **Durum Wheat**.

Durum Granular The product obtained from durum wheat by grinding to obtain semolina to which flour is added so that 7 to 20 percent passes through a number 100 U.S. sieve. It is used in macaroni and spaghetti. See **Durum Wheat**.

Durum Semolina See **Semolina**.

Durum Wheat The wheat obtained from the durum wheat kernel. It differs from other hard wheats in that the starch swelling capacity is greater and the gluten has different characteristics which result in tough, elastic doughs. As compared to hard wheat dough, it can be extruded through a small hole at lower pressure and in breads results in lower loaf volume. It is used almost exclusively in macaroni and spaghetti products because it is easily processed to produce a smooth, mechanically strong product of desired color which when cooked will maintain its shape and be of firm consistency. Products derived from the wheat include durum flour, durum granular, and durum semolina.

E

EDTA The abbreviation for ethylenediaminetetraacetate, a sequestrant and chelating agent that functions in water but not fats and oils. It is used to control the reaction of trace metals with some organic and inorganic components to prevent deterioration of color, texture, and development of precipitates, as well as to prevent oxidation which results in rancidity. The reactive sites of the metal ions are blocked, which prevents their normal reactions. The most common interfering metal ions in food products are iron and copper. It can be used in combination with the antioxidants BHT and propyl gallate. It is used in margarine, mayonnaise, and spreads to prevent the vegetable oil from going rancid. It is used in canned corn prior to retorting to prevent discoloration caused by trace quantities of copper, iron, and chromium. It also inhibits copper-catalyzed oxidation of ascorbic acid. It occurs as disodium calcium EDTA and disodium dihydrogen EDTA. Its use is approved in specified foods, with an average usage level being in the range of 100 to 300 parts per million.

Egg The hard-shelled reproductive body of poultry. The shell is largely composed of calcium carbonate, and represents approximately 11 percent of its total weight. Inside the shell are the shell membranes, which are principally protein. The yolk, which represents approximately 31 percent of the egg's weight, contains protein, fat, and all the known vitamins except vitamin C. Most of the egg's calories come from the yolk. The egg white is protein and represents approximately 58 percent of the weight. The white does not appear white in color until beaten or cooked. There is a thick and thin white, which consists mainly of ovalbumin, conalbumin, ovoglobulin, ovomucoid, and ovomucin. Eggs are used whole, as egg white, as yolk, or any combination thereof. They are used for coagulation, foam formation, emulsification, nutrition, flavor, and color.

Egg Albumen The protein fraction of egg, which is also termed egg white. It represents approximately 65 percent of the edible egg and is composed of approximately 87 percent water, 11 percent protein, and 1 percent carbohydrate. It provides a source of protein and provides foam upon whipping. It is used in meringues, cakes, and desserts.

Egg White See **Egg Albumen**.

Egg Yolk The yellow portion of the egg, representing approximately 35 percent of the edible egg. It is composed of approximately 49 percent water, 16 percent protein, 32 percent fat, and trace carbohydrate. It is used as an emulsifier in mayonnaise, salad dressing, and cream puffs. It is also used as a source of color.

Emulsifiers and Emulsifier Salts Substances which reduce the surface tension between two immiscible phases at their interface, allowing them to become miscible. The interface can be between two liquids, a liquid and a gas, or a liquid and a solid. Most emulsions involve water and oil or fat as the two immiscible phases, one being dispersed as finite globules in the other. The liquid as globules is referred to as the dispersed or internal phase, while the medium in which they are suspended as the continuous or external phase. There are two types of oil/water emulsions depending on the composition of the phases. In an oil-in-water emulsion such as milk and mayonnaise, the water is the external phase and the oil is the internal phase. In a water-in-oil emulsion such as butter, the oil is the external phase and the water is the internal phase. Emulsifiers have the following major functions:

1. Complexing—reaction with starch in bakery products which retards the crystallization of the starch, thus retarding the firming of the crumb which is associated with staling;
2. Dispersing—the reduction of interfacial tension which creates an intimate mixture of two liquids that normally are immiscible, example being oil-in-water emulsions such as salad dressing;
3. Crystallization control—control of crystallization in sugar and fat systems, i.e., chocolate, where it allows for brighter initial gloss and prevention of solidified fat on the surface;
4. Wetting—allows the surface to be more attracted to water, such as powders, i.e., coffee whitener, in which the addition of surfactant aids

the dispersion of the powder in the liquid without lumping on the surface; and

5. Lubricating—functions as a lubricant, such as in caramels, by reducing their tendency to stick to cutting knives, wrappers, and teeth.

Enriched Bleached Flour Flour that has been whitened by removal of the yellow pigments and fortified with vitamins and minerals. The added vitamins are thiamin, riboflavin, niacin, or niacinamide, and may include vitamin D. The minerals are iron and may include calcium. It is used in baked goods.

Entire Wheat Flour See **Whole Wheat Flour**.

Epsom Salt See **Magnesium Sulfate**.

Ergosterol A steroid alcohol that when irradiated with ultraviolet light yields calciferol (Vitamin D_2). Irradiated ergosterol is added to milk for vitamin D fortification.

Erythorbic Acid A food preservative that is a strong reducing agent (oxygen accepting) which functions similarly to antioxidants. In the dry crystalline state it is nonreactive, but in water solutions it reacts readily with atmospheric oxygen and other oxidizing agents, making it valuable as an antioxidant. During preparation, dissolving and mixing should incorporate a minimum amount of air and storage should be at cool temperatures. It has a solubility of 43 g per 100 ml of water at 25°C. One part erythorbic acid is equivalent to 1 part ascorbic acid and equivalent to 1 part sodium erythorbate. It is used to control oxidative color and flavor deterioration in fruits at 150 to 200 parts per million. It is used in meat curing to speed and control the nitrite curing reaction and prolong the color of cured meat at levels of 0.05 percent.

Erythrosine See **FD&C Red #3**.

Ethoxylated Mono- and Diglycerides An emulsifier prepared by the glycerolysis of edible vegetable fats and reacting with ethylene oxide. It is hydrophilic, being soluble in water and partially soluble in oil. It contributes to freeze-thaw stability and overrun in whipped toppings. It functions as a dough conditioner/emulsifier in baked goods and as an emulsifier in coffee whiteners, icings, and frozen desserts. Typical usage levels are 0.20 to 0.45 percent. It is also termed polyglycerate 60 and polyoxyethylene (20) mono- and diglycerides of fatty acids.

Ethoxyquin An antioxidant used in the preservation of color in chili powder, ground chili, and paprika.

Ethyl Acrylate A flavoring agent that is a clear, colorless liquid. Its odor is fruity, harsh, penetrating and lachrymatous (causes tears). It is sparingly soluble in water and miscible in alcohol and ether, and is obtained by chemical synthesis.

2-Ethylbutyric Acid A flavoring agent that is a clear liquid, colorless, with a rancid odor. It is miscible in alcohol and ether, sparingly soluble in water, and is obtained by chemical synthesis.

Ethyl Cellulose Used as a binder and filler in dry vitamin preparations, as a component of protective coatings for vitamin and mineral tablets, and as a fixative in flavoring compounds. It is a cellulose ether containing ethyoxy groups attached by an ether linkage and containing an anhydrous basis of not more than 2.6 ethoxy groups per anhydroglucose unit.

Ethyl Crotonate A synthetic flavoring agent that is a moderately stable, colorless to light yellow liquid of sharp winey note. It should be stored in glass, tin, or resin-lined containers. It is used in fruit flavors for application in baked goods, beverages, and candy at 2 to 7 parts per million.

Ethylene Oxide Polymer Foam stabilizer in fermented malt beverages which is the polymer of ethylene oxide. It is used at a level not to exceed 300 parts per million by weight of the fermented malt beverage. The label of the additive bears directions for use to insure compliance with the legal limit.

Ethylenediaminetetraacetate See **EDTA**.

Ethyl Formate A flavoring agent that occurs naturally in some plant oils, fruits, and juices but does not occur naturally in the animal kingdom. It is used in food in a maximum level, as served, of 0.05 percent in baked goods; 0.04 percent in chewing gum, hard candy, and soft candy; 0.02 percent in frozen dairy desserts; 0.03 percent in gelatins, puddings, and fillings; and 0.01 percent in all other food categories. It is an ester of formic acid and is prepared by esterification of formic acid with ethyl alcohol or by distillation of ethyl acetate and formic acid in the presence of concentrated sulfuric acid.

Ethyl-2,4-Hexadienoate See **Ethyl Sorbate**.

Ethyl Isobutyrate A synthetic flavoring agent that is a stable, colorless liquid of dry, fruity odor. It should be stored in tin, glass, or resin lined drums. It is used to give fruity effects to flavors for applications in candy, baked goods, and beverages at 10 to 100 parts per million.

Ethyl Lactate A solvent manufactured from L (+) lactic acid which is miscible in water and most organic solvents and is cleared for use as a flavoring agent. It is a naturally occurring constituent of California and Spanish sherries. It is a clear, colorless, nontoxic liquid of low volatility, having a pH of 7 to 7.5. It is used as a food and beverage flavoring agent.

Ethyl Maltol A flavoring agent that is a white, crystalline, powder. It has a unique odor and a sweet taste that resembles fruit. MP = 90°C (194°F). It is sparingly soluble in water and propylene glycol and soluble in alcohol and chloroform. It is obtained by chemical synthesis.

Ethyl-Methyl-Phenyl-Glycidate A synthetic flavoring agent that is a glycidic acid ester. It is a colorless to pale yellow liquid with a strong fruit odor suggestive of strawberries. It is unstable to alkali and moderately stable to weak organic acids. It should be stored in glass, tin, or aluminum containers. It is soluble in fixed oils and in propylene glycol. It is used in flavors for strawberry note and has application in candy, beverages, and ice cream at 6 to 20 parts per million. It is also termed aldehyde C-16.

Ethyl Nonanoate A synthetic flavoring agent that is a stable, colorless liquid of fruit cognac odor. It is practically insoluble in water and is miscible with alcohol and propylene glycol. It should be stored in glass or tin containers. It is used in flavors such as apple, pear, and cognac with applications in beverages, ice cream, candy, and alcohol beverages at 4 to 20 parts per million.

Ethyl Oxyhydrate See **Rum Ether**.

Ethyl Paraben See **Parabens**.

Ethyl Propionate A flavoring agent that is a transparent liquid, colorless, with an odor resembling rum. It is miscible in alcohol and propylene glycol, soluble in fixed oils, mineral oil, and alcohol, and sparingly soluble in water. It is obtained by chemical synthesis.

Ethyl Sorbate A synthetic flavoring agent that is a moderately stable, light yellow liquid of fruity odor. It should be stored in glass or tin containers.

It is used in flavors such as pineapple, papaya, and passion fruit with applications in ice cream, beverages, candy, and baked goods at 6 to 18 parts per million. It is also termed ethyl-2,4-hexadienoate.

Ethyl Vanillin A flavoring agent that is a synthetic vanilla flavor with approximately three and one-half times the flavoring power of vanillin. It has a solubility of 1 g in 100 ml of water at 50°C. It is used in ice cream, beverages, and baked goods.

Eugenol A flavoring obtained from clove oil and also found in carnation and cinnamon leaves. It is a stable, light yellow-green liquid of clove odor. It is slightly soluble in water and miscible in alcohol. It should be stored in glass or tin, avoiding iron containers. It is used in spice oils for application in condiments and meats at 100 to 200 parts per million and in baked goods and candy at approximately 30 parts per million.

Extract An alcohol or alcohol-water solution that contains a flavoring ingredient obtained from a spice or some other ingredient and which is used as a flavorant. It is used in baked goods, beverages, and ice cream.

Extract of Malted Barley and Corn See **Malted Cereal Syrup**.

F

Family Flour See **All-Purpose Flour**.

Farina Wheat, other than durum or red durum wheat, from which the bran and most of the germ has been removed. It is ground so that not more than 3 percent passes through a number 100 U.S. sieve.

Fast Green FCF See **FD&C Green #3**.

Fat Water-insoluble material of plant or animal origin, consisting predominantly of glyceryl esters of fatty acids (triglycerides). Fat ordinarily refers to triglycerides that are semisolid at room temperature. Fat in its liquid state is called oil.

Fatty Acids Aliphatic acids that may be saturated or unsaturated, consisting of a mixture of certain monobasic carboxylic acids and their associated fatty acids. Fatty acids plus glycerol result in a fat characterized by the fatty acid components. A fatty acid may be used as a lubricant, a binder, a food processing defoamer, and an emulsifier.

FD&C Blue #1 A colorant. It has a solubility in water of 20 g in 100 ml at 25°C. It has good pH stability with only slight fading after one week at pH 3, 7, 8 but is unstable in alkalis such as 10 percent sodium carbonate and 10 percent ammonium hydroxide. It has good stability in 10 percent sugar systems. It has fair stability to light, fair to poor stability to oxidation, and good stability to heat. It has a greenish-blue hue with excellent tinctorial strength. It is used with other primary colors to produce a variety of shades, for example, in combination with FD&C Yellow #5, it gives green. It has good compatibility with food components. It is used in candies, baked goods, soft drinks, and desserts. The common name is brilliant blue FCF.

FD&C Blue #2 A colorant. It has poor pH stability in that after one week at pH 3 and 5 it will appreciably fade, at pH 7 considerably fade, and at pH 8 fade completely. It is the least soluble of all food colors, with a solubility of 1.6 g in 100 ml of water at 25° C. Complete fading occurs in alkalis such as 10 percent sodium carbonate and 10 percent sodium hydroxide, with fading also occurring in 10 percent sugar systems. It has very poor light stability and oxidation stability, and moderate stability to heat. it has a deep blue hue with poor tinctorial strength. It is the only food color that has good resistance to reducing agents, but has very poor compatability with food components. The major use is in pet food, but it is also used in candies, confections, and baked goods. The common name is indigotine.

FD&C Green #3 A colorant. It has good pH stability, showing after one week a slight fade at pH 3, a very slight fade at pH 5 to 7, and slight fade and appreciably bluer color at pH 8. It has excellent solubility in water with a solubility of 20 g in 100 ml at 25°C. It has fair to good stability to light, poor stability to oxidation, and shows no appreciable change in 10 percent sugar systems. It has a bluish-green hue, with excellent tinctorial strength. It has good compatibility with food components and is occasionally used in cereals, soft drinks, beverages, and desserts. The common name is fast green FCF.

FD&C Red #3 A colorant. It is not recommended for use below pH 5.0, being insoluble at pH 3 to 5 but being stable at pH 7 and 8. It has a solubility in water of 9 g per 100 ml at 25°C. It has fair stability to oxidation and poor to fair stability to light, while having good stability in 10 percent sugar systems. It has exceptional clarity and brilliance, having a bluish-pink hue with very good tinctorial strength. It has poor compatibility with food components and is used in candies and confections as well as cherry dyeing. The common name is erythrosine.

FD&C Red #40 A colorant. It has good stability to pH changes from pH 3 to 8, showing no appreciable change. It has excellent solubility in water with a solubility of 22 g per 100 ml. at 25°C. It has very good stability to light, fair to poor stability to oxidation, good stability to heat, and shows no appreciable change in stability in 10 percent sugar systems. It has a yellowish-red hue and has a very good tinctorial strength. It has very good compatibility with food components and is used in beverages, desserts, candy, confections, cereals, and ice cream. The common name is Allura® red AC.

FD&C Yellow #5 A colorant. It has good stability to changes in pH, showing no appreciable change at pH 3 to 8. It has excellent solubility in water with a solubility of 20 g in 100 ml at 25°C. It has good stability to light and heat, fair stability to oxidation, and shows no appreciable change in 10 percent sugar systems. It has a lemon-yellow hue and has good tinctorial strength. It has moderate compatibility with food components and is used in beverages, baked goods, pet foods, desserts, candy, confections, cereal, and ice cream. The common name is tartrazine.

FD&C Yellow #6 A colorant. It has good stability to changes in pH, showing no appreciable change at pH 3 to 8. It has excellent solubility in water with a solubility of 19 g in 100 ml at 25°C. It has moderate stability to light, fair stability to oxidation, good stability to heat, and shows appreciable change in 10 percent sugar systems. It has a reddish-yellow hue and has good tinctorial strength. It has moderate compatibility with food components and is used in beverages, bakery goods, dessert confections, and ice cream. The common name is sunset yellow FCF.

Fennel A spice that is the dried, ripe fruit of the herb *Foeniculum vulgare* Mil. It is a seed with licorice flavor. It is used in meat, fish, and sauces as a seasoning.

Fenugreek The seed, usually in ground form, of the herb *Trigonella foenumgraecum*. It has a maple-like flavor and burnt sugar taste. It is used in curry powder, imitation maple flavor, chutney, and pickles.

Ferric Ammonium Citrate A nutrient and dietary supplement that is a source of iron, containing 17 percent iron.

Ferric Chloride A nutrient and dietary supplement that serves as a source of iron.

Ferric Citrate (Iron (III) Citrate) A nutrient supplement that is prepared from reaction of citric acid with ferric hydroxide. It is a compound of indefinite ratio of citric acid and iron. The ingredient may be used in infant formula.

Ferric Orthophosphate An inert white powder that is a source of iron and produces no discoloration or rancidity. It contains approximately 28 percent iron. It is used as a mineral supplement where rancidity is not a problem. It is used in frozen egg substitute, pasta products, and rice products.

Ferric Oxide A nutrient and dietary supplement that is a source of iron.

Ferric Phosphate (Ferric Orthophosphate, Iron (II) Phosphate) A nutrient supplement that is an odorless, yellowish-white to buff-colored powder and contains from one to four molecules of water of hydration. It is prepared by reaction of sodium phosphate with ferric chloride or ferric citrate.

Ferric Pyrophosphate (Iron (III) Pyrophosphate) A nutrient supplement, tan or yellowish white in color, prepared by reacting sodium pyrophosphate with ferric citrate. The ingredient may be used in infant formula.

Ferric Sulfate A nutrient and dietary supplement that is a source of iron.

Ferrous Ascorbate A nutrient supplement, blue-violet in color, containing 16 percent iron. It is a reaction product of ferrous hydroxide and ascorbic acid. May be used in infant formula.

Ferrous Carbonate A nutrient and dietary supplement that is a source of iron.

Ferrous Citrate A nutrient and dietary supplement that is a source of iron.

Ferrous Fumarate A reddish orange to red-brown powder that is a source of iron. It has high bioavailability and can be used in foods where the red color can be masked. It contains approximately 33 percent iron. It is used as a dietary supplement in breakfast cereals, poultry stuffing, enriched flour, and instant drinks.

Ferrous Gluconate A nutrient and dietary supplement that is a source of iron and a coloring adjunct. It is a yellowish gray to pale greenish yellow powder or granules with a burnt sugar odor. It has a solubility of 1 g in approximately 10 ml of water with slight heating. It is used by the pharmaceutical industry as an iron supplement in vitamin pills. It is used by olive growers to darken the olives to a uniform black color. It can function as an iron fortifier in corn and soy products, breakfast cereals, beverages, and dietary foods.

Ferrous Lactate The ferrous salt of L(+) lactic acid which is used as a source of iron. It is used in foods, such as infant foods, to supplement the needed dietary iron.

Ferrous Sulfate A nutrient and dietary supplement that is a source of iron. It is a white to grayish odorless powder. Ferrous sulfate heptahydrate contains approximately 20 percent iron, while ferrous sulfate dried contains approximately 32 percent iron. It dissolves slowly in water and has high bioavailability. It can cause discoloration and rancidity. It is used for fortification of baking mixes. In the encapsulated form it does not react with lipids in cereal flours. It is used in infant foods, cereals, and pasta products.

Fish Protein Isolate A food supplement that consists principally of dried fish protein prepared from the edible portion of fish after removal of the heads, fins, tails, bones, scales, viscera, and intestinal contents. The additive is prepared by extraction with hexane and food-grade ethanol to remove fat and moisture.

Flavor Enhancers Substances added to supplement, enhance, or modify the original taste and/or aroma of a food, without imparting a characteristic taste or aroma of its own. See **Flavoring Agents and Adjuvants**.

Flavoring Agents and Adjuvants Substances added to impart or help impart a taste or aroma in food. They are classified into the major groups of spices, natural flavors, and artificial or synthetic flavors. Aliphatic, aromatic, and terpene compounds refer to synthetic chemicals and isolates from natural sources. This classification encompasses the largest group of flavoring materials. The flavors used are natural, artificial, or combinations and exist in liquid or dry form. General flavor types available include fruit, dairy, meat, vegetable, beverage, and liquor.

Flour The food prepared by grinding and bolting cleaned wheat, other than durum wheat and red durum wheat. The baking quality of the flour depends upon the type of wheat, milling process, and treatment applied after milling. Flours classified by process are straight, patent, and clear flour. Flours classified by usage are all-purpose, bread, cake, cracker, and pastry flour. Flours treated after milling include bleached, bromated, enriched bleached, instantized, phosphated, and self-rising flour. Flours from other grains are identified according to the grain source, for example, soy flour. See specific flour.

Flour Treating Agents Substances added to milled flour, at the mill, to improve its color and/or baking qualities, including bleaching and maturing agents.

Foaming Agents See **Surface-Active Agents**.

Folacin See **Folic Acid**.

Folic Acid A water-soluble B-complex vitamin that aids in the formation of red blood cells, prevents certain anemias, and is essential in normal metabolism. High-temperature processing affects its stability. It is best stored at lower than room temperatures. It is also termed folacin. It is found in liver, nuts, and green vegetables.

Food Starch, Modified See **Modified Starch**.

Formic Acid A flavoring substance that is liquid and colorless, and possesses a pungent odor. It is miscible in water, alcohol, ether, and glycerin, and is obtained by chemical synthesis or oxidation of methanol or formaldehyde.

Fructose A sweetener that is a monosaccharide found naturally in fresh fruit and honey. It is obtained by the inversion of sucrose by means of the enzyme invertase and by the isomerization of corn syrup. It is 130 to 180 in sweetness range as compared to sucrose at 100 and is very water soluble. It is used in baked goods because it reacts with amino acids to produce a browning reaction. It is used as a nutritive sweetener in low-calorie beverages. It is also termed levulose and fruit sugar.

Fructose Corn Syrup A sweetener that is an isomerized corn syrup derived from isomerization of glucose in the syrup to fructose by the enzyme isomerase. Varying levels of fructose syrup are available, being designated 42, 55, and 90 percent fructose. The 42 percent high-fructose corn syrup (HFCS) is a liquid mixture of dextrose, fructose, maltose, isomaltose, and higher saccharides, of which 42 percent is fructose (dry basis). The 55 percent and 90 percent HFCS are liquid mixtures of fructose, dextrose, and higher saccharides containing 55 percent and 90 percent fructose (dry basis), respectively. The range of relative sweetness as compared to sucrose at 100 is 42 percent HFCS: 90 to 95; 55 percent HFCS: 95 to 100; 90 percent HFCS: 100 to 130. HFCS is used in carbonated beverages, canned fruit, frozen desserts, and dairy drinks. It is also termed isomerized syrup, levulose-bearing syrup, and high-fructose corn syrup.

Fruit Sugar See **Fructose**.

Fumaric Acid An acidulant that is a nonhygroscopic, strong acid of poor solubility. It has a solubility of 0.63 g in 100 ml of distilled water at 25°C. It dissolves slowly in cold water, but if mixed with dioctyl sodium sulfosuccinate its solubility improves. The solubility rate also increases with smaller particle size. A quantity of 0.317 kg of fumaric acid can replace 0.453 kg of citric acid. It is used in dry mixes such as desserts, pie fillings, and candy. It is used in dry beverage mixes because it is storage-stable, free-flowing, and nonhygroscopic. It functions as a synergistic antioxidant with BHA and BHT in oil- and lard-base products. In gelatin desserts, it improves the flavor stability, and increases shelf life and gel strength.

Furcellaran A gum that is the extract of the red alga *Furcellaria fastigiata*. It swells in cold water and requires heating to 75° to 80°C for solubilization. It forms thermoreversible gels after heating and cooling and has properties between agar and carrageenan. It is also termed Danish agar. It is used in milk puddings, flans, jelly, jam, and gelled meat products.

G

Garlic A spice that is cloves of the herb *Allium sativum*. In its dehydrated form, the flavor enzyme is released only when in combination with water. It exists in powder form and also as salt, chips, and seasoning powder. It is used to flavor meats, vegetables, and sauces.

Garlic Salt A seasoning that is a mix of garlic powder and salt. It is used in sauces and breads.

Gelatin A protein that functions as a gelling agent. It is obtained from collagen derived from beef bones and calf skin (Type B) or pork skin (Type A). Type B is derived from alkali-treated tissue and has an isoelectric point between pH 4.7 and 5.0. Type A is derived from acid-treated tissue and has an isoelectric point between pH 7.0 and 9.0 It forms thermally reversible gels which set at 20°C and melt at 30°C. The gel strength is measured by means of a Bloom Gellometer and ranges from 50 to 300 with a 250 Bloom being the most common. It is used in desserts at 8 to 10 percent of the dry weight, in yogurt at 0.3 to 0.5 percent, in ham coatings at 2 to 3 percent, and in confectionery and capsules at 1.5 to 2.5 percent.

Gellan gum A polysaccharide gum that reacts with mono-valent and di-valent salts to form gels. It is used in bakery fillings, icings, food gels, and spice adhesion. It is produced from *Pseudomonas elodea* by a pure culture fermentation process and purified by recovery with isopropyl alcohol. It is composed of tetrasaccharide repeat units, each containing one molecule of rhamnose and glucuronic acid and two molecures of glucose.

Geranyl Isovalerate A synthetic flavoring agent that is a moderately stable, light yellow liquid of fruity odor. It should be stored in glass or tin containers. It is used in fruit flavors such as apple or pear with applications in beverages, ice cream, candy, and baked goods at 4 to 11 parts per million.

Geranyl Phenylacetate A flavoring agent that is a yellow liquid with an odor resembling honey and roses. Miscible in alcohol, chloroform, and ether, and insoluble in water, it may contain other isomeric and closely-related terpenic esters. It is obtained by chemical synthesis.

Ghatti A gum that is a plant exudate obtained from the *Anogeissus latifolia* tree. The gum is formed as a protective sealant when the bark is damaged. It forms viscous mixtures in water at concentrations of 5 percent or greater. Only about 90 percent of the gum is actually soluble in water and has a pH of 4.5. It has similar uses as gum arabic. It is also termed Indian gum. It is used in buttered syrup and as a stabilizer for emulsions.

Ghee See **Butter Oil**.

Ginger A spice that is the dried and peeled rhizome of the ginger plant, *Zingiber officinale*. The fragrance ranges from pungent to piquant at once; the flavor can be sharp or cooling depending on the food with which it is used. Fresh (green) ginger is obtained from the cleaned, peeled, and cured rhizome; dried ginger is the fresh product which has been cured and ground for spice. It is used in desserts, meats, sauces, relishes, baked goods, and beverages.

Glacial Acetic Acid An acidulant that is a clear, colorless liquid which has an acid taste when diluted with water. It is 99.5 percent or higher in purity and crystallizes at 17°C. It is used in salad dressings in a diluted form to provide the required acetic acid. It is used as a preservative, acidulant, and flavoring agent. It is also termed acetic acid, glacial.

Gluconic Acid An acidulant that is a mild organic acid which is the hydrolyzed form of glucono-delta-lactone. It is prepared by the fermentation of dextrose, whereby the physiological D-form is produced. It is soluble in water with a solubility of 100 g per 100 ml at 20°C. It has a mild taste and at 1 percent has a pH of 2.8. It functions as an antioxidant and enhances the function of other antioxidants. In beverages, syrups, and wine, it can eliminate calcium turbidities. It is used as a leavening component in cake mixes, and as an acid component in dry-mix desserts and dry beverage mixes.

Glucono-Delta-Lactone An acidulant, abbreviated GDL. It hydrolyzes to form gluconic acid in water solution and thereby creates the desired pH. The rate of acid formation is affected by temperature, concentration, and the pH of the solution. It has low acid release at room temperature and accelerated conversion into gluconic acid at high temperatures. It is readily soluble with a solubility of 59 g in 100 ml of water at 20°C. It is used in chemical leavening and is used in sausages and frankfurters. It is an acidulant in dessert mixes.

Glucose See **Dextrose**.

Glucose Syrup See **Corn Syrup**.

Glutamic Acid An amino acid that is a white crystalline powder of slight solubility in water. The salt is monosodium glutamate (MSG) which functions as a flavor enhancer in meats. It also is a nutrient, dietary supplement, and salt substitute.

Glutamic Acid Hydrochloride A flavoring, salt substitute that is soluble in water and very slightly soluble in alcohol and ether. It is obtained by chemical synthesis.

Gluten A protein complex formed when water is kneaded with wheat flour which brings about the removal of a large portion of the starch. It forms the elastic framework of dough, entrapping the gas produced by the fermentation of leavening action which results in a risen dough of desired loaf volume and structure. Gliadin is of lower molecular weight and provides extensibility as compared to glutenin, which is of higher molecular weight and contributes elasticity. Gluten is available as wheat gluten, corn gluten, and zein. Vital wheat gluten is the most widely used. See **Wheat Gluten**.

Gluten Flour See **Gluten**.

Glyceral Tribenzoate A clouding agent used in beverage emulsions.

Glycerin A polyol (polyhydric alcohol) that functions as a humectant, crystallization modifier, and plasticizer. It is a bittersweet liquid which has a high solubility of 71 g per 100 g of water at 25°C. It is a fair oil solvent and has a medium to high hygroscopicity. It is used to maintain a certain moisture content to prevent the drying-out of foods. It is used in confections to maintain the initial level of crystallization of the soft sugar. It also functions as a flavor solvent. Applications include marshmallows, candy, and baked goods.

Glycerol See **Glycerin**.

Glycerol ester A density adjuster prepared from glycerol of non-animal sources and refined wood rosin of pine trees. It is used to adjust the specific gravity of the citrus oil or oil phase to be similar to the specific gravity of the beverage emulsion and thus prevent the oil from rising or settling in the finished beverage. It also imparts some cloudiness. It is soluble in aromatic and petroleum hydrocarbons, terpenes, esters, ketones, citrus, and essential oils. It is used in lemon and orange drinks and also as a masticatory substance in chewing gum base. It is technically termed glyceryl abietate and is also called glycerol dihydroabietate.

Glyceryl-Lacto Esters of Fatty Acids Lipophilic emulsifiers that are the lactic acid esters of mono- and diglycerides. They are made by the reaction of mono- and diglycerides or propylene glycol ester with lactic acid, resulting in a compound with more surface activity and slightly more hydrophilicity than the regular mono- and diglycerides. They are used as emulsifiers, plasticizers, and promoters of starch gelatinization. They are used where aeration is required, such as in toppings, cakes, and icings, at levels necessary to obtain the technical effect.

Glyceryl-Lacto-Stearate An emulsifier that is a glyceryl-lacto ester of fatty acids. It is a monoglyceride esterified with lactic acid which increases the hydrophilicity of the emulsifier. It is used in whipped vegetable toppings, shortenings, cake mixes, and chocolate coating.

Glyceryl Monolaurate A monoglyceride emulsifier produced by the esterification of glycerine and lauric acid. It has a melting point of 56°C, a maximum iodine value of 0.5, and a saponification value of 200 to 206. In a highly purified form, it shows antimicrobial properties against microorganisms with the exception of gram-negative organisms. It is effective against gram-negative organisms when formulated with BHA or EDTA. It is used in baked goods, whipped toppings, frosting, glazes, and cheese products.

Glyceryl Monooleate A flavoring agent that is prepared by esterification of commercial oleic acid that is derived either from edible sources or from tall oil fatty acids. It contains glyceryl monooleate and glyceryl esters of fatty acids present in commercial oleic acid. The ingredient is also used as an adjuvant and as a solvent and vehicle.

Glyceryl Monostearate Glyceryl monostearate, also known as monostearin, is a mixture of variable proportions of glyceryl monostearate, glyceryl monopalmitate, and glyceryl esters of fatty acids present in commercial stearic acid. Glyceryl monostearate is prepared by glycerolysis of certain fats or oils that are derived from edible sources or by esterification, with glycerin, of stearic acid that is derived from edible sources.

Glyceryl Triacetate A colorless, oily liquid of slight fatty odor and bitter taste. It is soluble with water and is miscible with alcohol and ether. It functions in foods as a humectant and solvent. It is also termed triacetin.

Glyceryl Tristearate A formulation aid, lubricant, and release agent, prepared by reacting stearic acid with glycerol in the presence of a suitable catalyst. The additive is used as a crystallization accelerator in cocoa products; a formulation aid in confections; a formulation in fats and oils; and a winterization and fractionation aid in fat and oil processing.

Glycine A nonessential amino acid that functions as a nutrient and dietary supplement. It has a solubility of 1 g in 4 ml of water and is abundant in collagen. It is used to mask the bitter aftertaste of saccharin, for example, in artificially sweetened soft drinks. It retards rancidity in fat.

Glycyrrhizin A flavorant and foaming agent derived from the licorice root *Glycyrrhiza glabra*. It is 50 to 100 times as sweet as sugar, is soluble in water, and has a licorice taste. It has foaming and emulsifying properties in water, being used in cocktail mixes and soft drinks. It is used as a flavorant in bacon and imitation whipped products. It is synergistic with sugar, the sweetness being amplified to 100 times that of cane sugar alone. It is used as a sweetener in sugar-free chewing gum and low-fat sugar-free frozen desserts. It is also termed ammoniated glycyrrhizin.

Golden Apple Seed See **Quince Seed**.

Graham Flour Another term for whole wheat flour.

Graham's Salt See **Sodium Polyphosphate** and **Sodium Tetrametaphosphate**.

Grain Vinegar An acidulant made by the acetous fermentation of dilute distilled alcohol, containing not less than 4 g of acetic acid per 100 ml at 20°C. It is used in mayonnaise, salad dressing, sauces, and catsup. It is also termed distilled vinegar and spirit vinegar.

Granulated Sugar See **Sugar**.

Grape Color Extract An aqueous solution of anthocyanin grape pigment made from Concord grapes, or a dehydrated water-soluble powder prepared from the aqueous solution. It contains the common components of the grape juice, but not in the same proportions. It has a red color pigment, with greatest color stability below pH 4.5. The color is stable in the presence of light and some heat. The color intensity increases as the pH declines. It is used at the 0.05 to 0.8 percent range. It may be used for coloring nonbeverage foods.

Grape Seed Oil The oil obtained from grape seeds which contain an average of 15 percent oil. It is used as a drying oil with seeded raisins to improve their appearance and to prevent sticking. It is also termed raisin seed oil.

Grape Skin Extract A natural red colorant with a high concentration of red anthocyanic pigments which provide its physicochemical properties. These pigments are responsible for the red, purple, violet, and blue hues of flowers and fruits. It is prepared by aqueous extraction of the fresh seedless marc remaining after the grapes have been pressed in the production of grape juice and wine. It contains the common components of grape juice, but in different proportions. The color depends upon the medium and the pH. In an acid medium and up to pH 4.5 to 5.5, the color is violet and becomes blue at pH 6.5. It has excellent water solubility and fair heat, light, and chemical stability. It can be used in soft drinks at 0.2 to 0.4 percent, in candies at 0.5 to 1.5 percent.

Guaiacol A precursor of vanillin and santalidol (a synthetic sandalwood fragrance). It is obtained from wood tar by the destructive distillation of hardwood, by the distillation of the phenol fraction of coal tar, or through the use of o-dichlorobenzene. It is processed to yield vanillin.

Guar A gum that is a galactomannan obtained from the seed kernel of the guar plant *Cyamopsis tetragonoloba*. It is dispersible in cold water to form viscous sols which upon heating will develop additional viscosity. A 1 percent solution has a viscosity range of 2000 to 3500 centipoises. It is a versatile thickener and stabilizer used in ice cream, baked goods, sauces, and beverages at use levels ranging from 0.1 to 1.0 percent. It is scientifically termed guaran.

Guaran See **Guar**.

Gum Base The component of chewing gum that is insoluble in water and remains after chewing. It is prepared by blending and heating several ingredients to include a masticatory substance of vegetable or synthetic origin such as chicle, crown gum, petroleum wax, lanolin, polyethlyene, polyvinyl acetate, or rubber, with a plasticizer such as paraffin and with antioxidants. The gum base is 15 to 30 percent of chewing gum, of which a sweetener is the principal ingredient.

Gum Arabic See **Ghatti**.

Gum Ghatti See **Ghatti**.

Gum Tragacanth See **Tragacanth**.

Gum Quince Seed See **Quince Seed**.

Gums, or Hydrocolloids Polysaccharides that function as water-control agents by increasing viscosity (resistance to flow) or by forming gels. Gums are classified by source according to the following principal groupings: plant exudates, which include arabic, tragacanth, karaya, and ghatti; seaweed extracts, which include agar, alginates, carrageenan, and furcellaran; plant seed gums, which include guar, locust bean, tamarind, psyllium, and quince; plant extracts, which include pectin and arabinogalactan; fermentation gums, which include xanthan gum and dextran; and cellulose derivatives, which include carboxymethyl cellulose, hydroxypropylmethyl cellulose, and microcrystalline cellulose. Gum derivatives include propylene glycol alginate and low-methoxyl pectin.

Heptanone (Methyl Amyl Ketone) A flavoring agent that is miscible in alcohol and ether, slightly soluble in water. It is obtained by chemical synthesis. This flavoring substance or its adjuvant may be safely used in food in the minimum quantity required to produce its intended flavor. It can be used alone or in combination with other legally approved flavoring substances or adjuvants.

Heptyl Cinnamate A synthetic flavoring agent that is a fairly stable, yellow liquid with a hyacinth odor. It should to stored in glass or tin containers. It is used to smooth out fruity flavors and has application in gelatins and puddings at approximately 20 parts per million and in candy, beverages, and ice cream at 2 to 6 parts per million.

Heptyl Formate A synthetic flavoring agent that is a moderately stable, colorless to light yellow liquid of fruity odor. It should be stored in glass or tin containers. It is used in fruit flavors such as apricot, pear, and plum with applications in beverages, ice cream, candy, and baked goods at 1 to 4 parts per million.

Heptyl Isobutyrate A synthetic flavoring agent that is a stable, colorless liquid of fruity odor. It should be stored in glass or tin containers. It is used in flavors for pineapple, pear, and orange with applications in beverages, ice cream, candy, and baked goods at 1 to 3 parts per million.

Heptyl Paraben A preservative and antimicrobial agent. It is very slightly soluble in water. It may be used in fermented malt beverages to inhibit microbial spoilage and is permitted in beer. It is also termed N-heptyl-para-hydroxybenzoate.

Hesperidin A flavoring agent that is a bioflavonoid found in citrus pulp. It has minor use as a flavorant.

High-Fructose Corn Syrup (HFCS) A sweetener that is an isomerized corn syrup derived from the isomerization of the glucose in the syrup to fructose by the enzyme isomerase. Varying concentrations of fructose syrup are available, designated 42, 55, and 90 percent fructose. The 42 percent high-fructose corn syrup (HFCS) is a liquid mixture of dextrose, fructose, maltose, isomaltose, and higher saccharides, of which 42 percent is fructose, dry basis. The 55 and 90 percent HFCS's are liquid mixtures of fructose, dextrose, and higher saccharides containing 55 and 90 percent fructose, dry basis, respectively. The range of relative sweetness as compared to sucrose at 100 is 42 percent HFCS: 90 to 95; 55 percent HFCS: 95 to 100; 90 percent HFCS: 100 to 130. It is used in carbonated beverages, canned fruit, frozen desserts, and dairy drinks. It is also termed isomerized syrup, levulose-bearing syrup, and fructose corn syrup.

Homogenized Milk Milk that has been mechanically treated to reduce the size of the fat globules such that after 48 hours of quiescent storage at about 7°C no visible cream separation occurs and the percentage of fat of the upper 100 ml in 946 ml of milk does not differ by more than 10 percent from the fat percentage of the remaining milk. Homogenization makes the milk more homogeneous but also decreases the heat stability of the milk proteins. It is used as a beverage and constituent of other food products. Practically all whole milk sold retail in the United States is homogenized.

Honey A sweetener that is a natural syrup. It is similar to invert sugar, with a small but variable excess of levulose (fructose). It is formed by the action of the enzyme honey invertase on nectar gathered by bees. The composition and flavor varies with the plant source of the nectar, processing, and storage. A typical composition is 41 percent fructose, 34 percent glucose, 18 percent water, and 2 percent sucrose with a pH of 3.8 to 4.2. It is used in foods as a sweetener.

Horseradish A spice, the granules obtained from the horseradish plant. The flavor is released with moisture. It has a hot flavor character and has good stability. It is used in sauces.

Hydrated Lime See **Calcium Hydroxide.**

Hydrochloric Acid An acid that is the aqueous solution of hydrogen chloride of varying concentrations. It is miscible with water and with alcohol. It is used as an acidulant and neutralizing agent.

Hydrogenated and Partially Hydrogenated Menhaden Oils Used as edible fats or oils, partially hydrogenated and hydrogenated menhaden oils are prepared by feeding hydrogen gas under pressure to a converter containing crude menhaden oil and a nickel catalyst. The reaction is begun at 150° to 160°C and after 1 hour the temperature is raised to 180°C until the desired degree of hydrogenation is reached. Hydrogenated menhaden oil is fully hydrogenated. If the fat or oil is fully hydrogenated, the name used on the label of a product containing it includes the term "hydrogenated"; if the oil is partially hydrogenated, the label includes the term "partially hydrogenated."

Hydrogenated Vegetable Oil Oil that has been hydrogenated to modify the texture from a liquid to a semisolid or solid. The hydrogenization, which is the chemical addition of hydrogen, raises the melting point and converts the oil to a more desirable texture and consistency. It is used in farinaceous foods, confectionary, and desserts.

Hydrolyzed Cereal Solids These are maltodextrins of low DE (Dextrose Equivalent). They function as anticaking agents, bodying agents, carriers, and crystallization inhibitors and are used in dry mixes and desserts.

Hydrolyzed Protein See **Hydrolyzed Vegetable Protein.**

Hydrolyzed Vegetable Protein (HVP) A flavor enhancer obtained from vegetable proteins such as wheat gluten, corn gluten, defatted soy flour, and defatted cottonseed. The proteins are hydrolyzed into their component amino acids after which the reaction mixture is neutralized with sodium carbonate and refined. The refined liquid HVP consists of amino acids, monosodium glutamate, amino acid derivatives, salt, and water. After being stored for several months, the liquid HVP is concentrated into a paste, dried, and ground. A typical dried HVP consists of 40 to 45 percent salt, which is generated during the neutralization process and serves to enhance the mouth feel of the HVP and provide preservation properties. It normally contains 9 to 12 percent monosodium glutamate and the remaining fraction consists of flavor solids. There are two basic types: pale HVP, which functions as a flavor enhancer with delicate spray flavors used in cream-type soups and sauces, and poultry; and dark HVP, which functions as a flavor donor with strong meaty flavors used in stews and broths. HVP is stable under varying processing conditions. It is used to improve flavors in soups, dressings, meats, snack foods, and crackers.

Hydroxylated Lecithin An emulsifier and clouding agent that is a modified crude lecithin of improved water dispersibility. It is manufactured by treating soybean lecithin with peroxide to increase the hydrophilic properties of lecithin. It is partially soluble in water but hydrates readily to form emulsions. It is used in bakery products because it has an apparent synergy with mono- and diglycerides. It is also used in dry-mixed beverages and margarine. It is also termed hydroxylated soybean lecithin.

Hydroxylated Soybean Lecithin See **Hydroxylated Lecithin.**

4-Hydroxymethyl-2,6-Di-Tert-Butylphenol An antioxidant used alone or in combination with other permitted antioxidants. The total amount of all antioxidants added to such food must not exceed 0.02 percent of the oil or fat content of the food, including the essential (volatile) oil content of the food.

Hydroxypropyl Cellulose A gum that is nonionic water-soluble cellulose, obtained from the reaction of alkali cellulose with propylene oxide at high temperatures and pressures. It is soluble in water below 40°C, is precipitated as a floc between 40° and 45°C, and is insoluble above 45°C. The precipitation is reversible with the original viscosity being restored upon cooling below 40°C and stirring. It is used in whipped toppings as a stabilizing and foaming aid; in edible food coatings as a glaze and oil/oxygen barrier; and in fabricated foods as a binder. Typical usage level is 0.05 to 1.0 percent.

Hydroxyproply Methylcellulose A gum formed by the reaction of propylene oxide and methyl chloride with alkali cellulose. It will gel as the temperature is increased in heating and upon cooling will liquefy. The gel temperature ranges from 60 to 90°C, forming semi-firm to mushy gels. It is used in bakery goods, dressings, breaded foods, and salad dressing mix for syneresis control, texture, and to provide hot viscosity. Usage level ranges from 0.05 to 1.0 percent.

I

Indian Gum See **Ghatti.**

Indigotine See **FD&C Blue #2.**

Indole A flavoring agent that is a white, flaky crystalline product. It has an unpleasant odor when concentrated and a flowery odor when diluted. It is soluble in most fixed oils and propylene glycol and insoluble in glycerin and mineral oil. It is obtained from decomposition of a protein.

Instantized Flour A flour made by a milling or agglomerating procedure which makes it readily pourable, providing convenience.

Invert Sugar A sweetener that is a mixture of equal weights of dextrose (glucose) and levulose (fructose). It is more soluble than sucrose and has higher moisture-retaining properties because of the fructose content. It resists crystallization. It is used in candy and icings because it is sweeter, more soluble, and crystallizes less readily than sucrose.

Invert Sugar Syrup A sweetener produced by an inversion process. It is produced by solubilizing sucrose in water followed by hydrolization to a mixture of dextrose and fructose using acids, invertase enzyme, or ion exchange resins to catalyze the reaction. Several invert syrups are obtained, such as medium invert consisting of 50 percent sucrose, 25 percent dextrose, 25 percent fructose; and total invert consisting of 3 to 5 percent sucrose, 48 percent dextrose, and 47 percent fructose. It has improved microbiological stability because of its high solids content and it is used in soft drinks. It is also termed sugar syrup, invert.

Iodine A halogen element extracted from Chilean nitrate-bearing earth or from seaweed. It functions by its presence in the thyroid hormones. Iodine deficiency is associated with goiter. Sources are potassium and cuprous iodide and potassium and calcium iodate, of which the iodate form is preferred because of better stability. It is used as a food supplement.

Irish Moss See **Carrageenan.**

Iron A mineral used in food fortification that is necessary for the prevention of anemia, which reduces the hemoglobin concentration and thus the amount of oxygen delivered to the tissues. Sources include ferric ammonium sulfate, chloride, fructose, glycerophosphate, nitrate, phosphate, pyrophosphate and ferrous ammonium sulfate, citrate, sulfate, and sodium iron EDTA. The ferric form (Fe^{+3}) is iron in the highest valence state and the ferrous form (Fe^{+2}) is iron in a lower valence state. The iron source should not discolor or add taste and should be stable. Iron powders produce low discoloration and rancidity. It is used for fortification in flour, baked goods, pasta, and cereal products.

Iron Ammonium Citrate An anticaking agent used in salt.

Iron, elemental. A nutrient supplement, metallic iron is obtained by any of the following processes: reduced iron, electroytic iron, and carbonyl iron.

Iron-Choline Citrate Complex This special dietary additive is made by reacting approximately equimolecular quantities of ferric hydroxide, choline, and citric acid, and is used as a source of iron.

Iron Oxide A trace mineral used as a pigment and colorant. It is used to color pet food.

Iron, Reduced Iron in a lower valence state, such as the ferrous form (Fe^{+2}). It is used in dry-mix oatmeal.

Isoamyl Acetoacetate A synthetic flavoring agent that is a stable, colorless liquid of light green leaf-fruity odor. It should be stored in glass or tin containers. It is used in currant and berry flavors for applications in beverages, candy, and ice cream at 5 to 15 parts per million.

Isoamyl Butyrate A synthetic flavoring agent that is a stable, colorless liquid of strong fruity odor. It is usually prepared by esterification of isoamyl alcohols with butyric acid. It is soluble in most fixed oils and mineral oil and is insoluble in glycerin and propylene glycol. Storage should be in glass, tin, or resin-lined containers. It is used in fruit flavors such as pineapple, raspberry, and strawberry and has application in dessert gels, puddings, and baked goods at 50 to 60 parts per million.

Isoamyl Formate A synthetic flavoring agent that is a moderately stable, colorless to light yellow liquid of pungent pear-plum odor, being soluble in most fixed oils, mineral oil, and propylene glycol. Storage should be in a glass or tin container. It is used in fruit flavors such as pear, plum, and peach for application in dessert gels, puddings, candy, and ice cream at 14 to 28 parts per million.

Isoamyl Hexanoate A synthetic flavoring agent that is a stable, colorless liquid of fruity odor. It is soluble in alcohol, fixed oils, and mineral oil. Storage should be in glass, tin, or resin-lined containers. It is used in fruit flavors such as banana and pineapple for applications in desserts, candy, and ice cream at 4 to 22 parts per million.

Isoascorbic Acid See **Ascorbic Acid.**

Isobutyl Acetate A flavoring agent that is a clear colorless liquid with a fruity odor resembling banana when diluted. It is soluble in alcohol, propylene glycol, most fixed oils, and mineral oil, and slightly soluble in water. It is obtained by synthesis.

Isobutyl Cinnamate A synthetic flavoring agent that is a stable, colorless to light yellow liquid of fruity odor. It is miscible with alcohol, chloroform, and ether but is practically insoluble in water. Storage should be in glass or tin-lined containers. It is used in fruit flavors such as cherry and prune with applications in beverages, ice cream, candy, and baked goods at 1 to 5 parts per million.

Isobutyl Formate A synthetic flavoring agent that is a stable, colorless liquid of fruity odor. Storage should be in glass or tin containers. It is used in fruit flavors such as pear, raspberry, and other berry flavors with applications in beverages, ice cream, candy, and baked goods at 2 to 18 parts per million.

Isobutyric Acid (Isopropylformic Acid) A flavoring agent that is a colorless liquid with a strong, penetrating odor, resembling butter. It is miscible in alcohol, propylene glycol, glycerin, mineral oil, and most fixed oils and soluble in water. It is obtained by chemical synthesis.

Isolated Soy Protein See **Soybean Protein Isolate.**

Isomerized Syrup See **High-Fructose Corn Syrup.**

Isopropyl Citrate An antioxidant that reacts with metal ions that might catalyze oxidative reactions and thus will prevent rancidity. It is made by reacting citric acid (not soluble in fats and oils) with isopropyl alcohol (which readily dissolves in oil) and thus enables the citrate to dissolve in oil. It is used in vegetable oils.

J

Juniper Berries Oil A flavoring agent that is a liquid which may be colorless, yellow or greenish in appearance. Its odor is characteristic with an aromatic, bitter taste. Storage is accompanied by polymerization. It is soluble in most fixed oiis and mineral oil, insoluble in glycerin and propylene glycol. It is obtained from dried ripe fruit of *Juniperus communis* L. var. *erecta* Pursh of the *Cupressaceae* family.

K

Karaya A gum, the dried exudate from the *Sterculia urens* tree which is native to India. It does not dissolve in water but swells to form a colloidal sol with a rate of hydration depending on mesh size. A 3 to 4 percent sol will result in a heavy gel and for higher concentrations the gum must be cooked under steam pressure to solubilize. It has a pH of 4.5 to 4.7. It functions as a binder and adhesive. It is used in baked goods, denture adhesives, toppings, and frozen desserts.

Kelp A brown seaweed that grows in ocean water. The principal commercial species include *Macrocystis pyrifera* and *Laminaria hyperboria.* It is a source of alginic acid, which is used to produce alginate gum which functions as a water control agent. It contains the trace minerals potassium, sodium, calcium, and iodine. It is used as a source of iodine, as a flavor enhancer, as a nutrient and dietary supplement, and as a source of alginates.

Kola Nut The seed of *Cola nitida* or other *Cola* species. The nut contains approximately 1.5 percent caffeine and is used in beverages and as an adjunct with other flavors.

L

Lactalbumin A milk protein obtained from whey by acidifying to pH 5.2, the isoelectric point, followed by coagulation by heat. It is not coagulated by rennin as in casein and is nonfunctional in its properties. It is used for nutritional purposes as a source of protein. It is used in cereals and breads where its relative inertness minimizes complications caused by other milk proteins during baking. It is also termed milk albuminate.

Lactase Enzyme Preparation from *Kluyveromyces lactis* An enzyme preparation used to convert lactose to glucose and galactose. It is derived from the nonpathogenic, nontoxicogenic yeast *Kluyveromyces lactis* (previously named *Saccharomyces lactis*), and contains the enzyme B-galactoside galactohydrase, which converts lactose to glucose and galactose. It is prepared from yeast that has been grown in a pure culture fermentation and by using materials that are generally recognized as safe or food additives that have been approved for this use. This ingredient is used in milk to produce lactase-treated milk, which contains less lactose than regular milk, or lactose-reduced milk, which contains at least 70 percent less lactose than regular milk.

Lactic Acid An acidulant that is a natural organic acid present in milk, meat, and beer, but is normally associated with milk. It is a syrupy liquid available as 50 and 88 percent aqueous solutions, and is miscible in water and alcohol. It is heat stable, nonvolatile, and has a smooth, milk acid taste. It functions as a flavor agent, preservative, and acidity adjuster in foods. It is used in Spanish olives to prevent spoilage and provide flavor, in dry egg powder to improve dispersion and whipping properties, in cheese spreads, and in salad dressing mixes.

Lacto Esters of Propylene Glycol See **Lactylated Fatty Acid Esters of Glycerol and Propylene Glycol.**

Lactoglobulin A protein that is a complex of closely related proteins known as beta-globulins obtained from the whey fraction of milk. It is crystallizable and heat-denaturable.

Lactose A disaccharide carbohydrate that occurs in mammalian milk except that of the whale and the hippopotamus. It is principally obtained as a cows' milk derivative. It is also termed milk sugar and it is a reducing sugar consisting of glucose and galactose. Its most common commercial form is alpha-monohydrate, with the beta-anhydride form available to a lesser extent. All forms in solution will equilibrate to a beta : alpha ratio of 62.25 : 37.75 at 0°C. It is about one-sixth as sweet as sugar and is less soluble. It functions as a flow agent, humectant, crystallization control agent, and sweetener. It is used in baked goods for flavor, browning, and tenderizing and in dry mixes as an anticaking agent.

Lactylated Fatty Acid Esters of Glycerol and Propylene Glycol An emulsifier made by the reaction of a propylene glycol ester with lactic acid. It has more surface activity and is slightly more hydrophilic than mono- and diglycerides. It is used mainly where aeration is required, such as in toppings, cake mixes, and icings. It is used at levels required to produce the intended effect, such as 0.6 percent in fluid whipped topping and 0.5 percent in coffee whitener.

Lactylic Esters of Fatty Acids An emulsifier that is mixed fatty acid esters of lactic acid and its polymers. It is dispersible in water and soluble in organic solvents and vegetable oils. It functions as a foaming agent in starch/protein systems and is used in puddings and coffee whiteners.

Larch See **Arabinogalactan.**

Lard A fat rendered from hogs, consisting principally of oleic and palmitic fatty acids. It has a Wiley melting point of 88° to 110°F. It is rapidly chilled, resulting in an opaque, firm consistency rather than a translucent, greasy appearance. It is used in cake mix.

Lauric Acid A fatty acid obtained from coconut oil and other vegetable fats. It is practically insoluble in water but is soluble in alcohol, chloroform, and ether. It functions as a lubricant, binder, and defoaming agent.

Leavening Agents Acidic agents that chemically react with alkaline sodium bicarbonate to produce carbon dioxide gas. This reaction is initiated by moisture and completed by heat as the prepared mixture is baked. The value of the leavening agent relates to the rate upon which carbon dioxide is released from sodium bicarbonate, the suitability of the release rate to the product, and the mixing-raising-baking cycle. Leavening agents include tartaric acid, monocalcium phosphate, sodium acid pyrophosphate, sodium aluminum phosphate, and acidic acid.

Lecithin An emulsifier that is a mixture of phosphatides which are typically surface-active. It is now commercially obtained from soybeans; previously it was obtained from egg yolk. It is used in margarine as an emulsifier and antispatter agent; in chocolate manufacture it controls flow properties by reducing viscosity and reducing the cocoa butter content from 3 to 5 percent; it is used as a wetting agent in cocoa powder, fillings, and beverage powders; an antisticking agent in griddling fat; and in baked goods to assist the shortening mix with other dough ingredients and to stabilize air cells. Typical usage levels range from 0.1 to 1.0 percent.

Lecithinated Soy Flour Soy flour to which lecithin is added. The lecithin contributes emulsification and pan release properties. It is used in breading, caked foods, and dough mixes.

Lemon Oil A flavoring agent that is the oil obtained from lemon fruit. It is used to impart lemon flavor and is used in reconstituted lemon juice.

Levulose See **Fructose.**

Levulose-Bearing Syrup See **High-Fructose Corn Syrup.**

Licorice A flavoring agent made from dried root portions of *Glycyrrhiza glabra*. The obtainable forms are licorice root, licorice extract powder, and licorice extract. The extract is used in candy, baked goods, and beverages; the major licorice use is in tobacco.

Lime See **Calcium Oxide.**

Limestone See **Calcium Carbonate.**

Limonene An antioxidant and flavoring agent that occurs in lemons, oranges, and pineapple juice, being obtained from the oils. It is a colorless liquid which is insoluble in water and propylene glycol, very slightly soluble in glycerine, and miscible with alcohol, most fixed oils, and mineral oil. It prevents or delays enzymatic browning-type oxidation.

Linalyl Isobutyrate (3,7-Dimethyl-2,6-Octadien-3-yl Isobutyrate) A flavoring agent that is a liquid, slightly yellow in color with a fruity odor. It is miscible in alcohol, ether, and chloroform, and insoluble in water. It is obtained by chemical synthesis.

Locust Bean Gum A gum that is a galactomannan obtained from the plant seed from the locust bean tree known as *Ceratonia siligua.* Its properties include swelling partially in cold water but requiring heating to approximately 82°C for complete solubility. It provides high viscosity, forms gels with xanthan gum upon heating and cooling of the solution, and functions as a water binder. It can made agar or carrageenan gels more elastic. Its uses include processed cheese, ice cream, bakery products, soups, and pies. Typical usage level is 0.1 to 1.0 percent. It is also called carob gum or Saint John's bread, and is scientifically called carubin.

Low-Methoxyl Pectin A gum derived from pectinic acid. It differs from pectin in having a lower degree of methylation, less than 50 percent. It is also not as sensitive to pH and does not require sugar for gel formation. It forms thermally reversible gels with calcium salts and boiling may be required for solubility if the methoxyl content is low. It is used in low-calorie jellies at levels of 0.8 to 1.4 percent, in dessert and bakery jellies at levels of 1 to 1.5 percent, and in fruit gels at levels of 0.5 to 1.4 percent.

M

Mace A spice that is the aril or skin covering of the nutmeg *Myristica fragrans* Houtt. It is more pungent in flavor than nutmeg. The whole mace is used in cooked fruit, pickles, and preserves, while ground mace is used in breads, cakes, chocolate pudding, and fruit salad.

Magnesium A metallic element that is involved in certain bodily functions. Sources of magnesium include magnesium chloride and magnesium oxide. It functions as a nutrient and dietary supplement.

Magnesium Carbonate An anticaking agent and general purpose food additive. It is practically insoluble in water but is more soluble in water containing carbon dioxide. It imparts a slightly alkaline reaction to the water. It is used as an alkali in sour cream, butter, and canned peas. It is used as an anticaking agent in table salt and dry mixes. It assists in providing clarity in algin gels and functions as a filler in dental impression materials.

Magnesium Caseinate The magnesium form of caseinate which is marginal in functionality as compared to other forms of caseinates. It can be used in bakery goods, drinks, and dietary applications. See **Caseinates.**

Magnesium Chloride A source of magnesium, a color-retention agent, and firming agent. It exists as colorless flakes or crystals and is very soluble in water.

Magnesium Hydroxide An alkali that is a general purpose food additive. It exists as a white powder and has poor solubility in water and in alcohol.

In frozen desserts it will increase the tendency for fat globules to clump, which results in an increase in dryness. It reacts with triglycerides in fatty acids to form soaps. It also functions as a drying agent in foods.

Magnesium Laurate The magnesium salt of lauric acid which functions as a binder, emulsifier, and anticaking agent.

Magnesium Myristate The magnesium salt of myristic acid which functions as a binder, emulsifier, and anticaking agent.

Magnesium Oleate The magnesium salt of oleic acid which functions as a binder, emulsifier, and anticaking agent.

Magnesium Oxide A source of magnesium which functions as a nutrient and dietary supplement. It exists as a bulky white powder termed light magnesium oxide or as a dense white powder known as heavy magnesium oxide. It is practically insoluble in water and is insoluble in alcohol.

Magnesium Palmitate The magnesium salt of palmitic acid which functions as a binder, emulsifier, and anticaking agent.

Magnesium Silicate A white powder that is insoluble in water and functions as an anticaking agent. It is used in salt. It also is a processing aid and adsorbent which functions as an anticaking agent and remover of undesirable proteins during filtration. It is insoluble and a 10 percent slurry has a pH of approximately 7.0. It aids in the processing of beverages, food products, and pharmaceuticals by removing protein/tannin complex constituents through surface area and adsorptive effects.

Magnesium Stearate The magnesium salt of stearic acid which functions as a lubricant, binder, emulsifier, and anticaking agent. It is a white powder that is insoluble in water. It is used as a lubricant or die release in tableting pressed candies and is also used in sugarless gum and mints.

Magnesium Sulfate A nutrient and dietary supplement which is also termed Epsom salt. It is soluble in water and its solutions are neutral. It exists as crystals with a cooling, saline, bitter taste.

Maize Meal The meal obtained by grinding maize (Indian corn).

Maize Starch See **Cornstarch.**

Malic Acid An acidulant that is the predominant acid in apples. It exists as white crystalline powder or granules and is considered hygroscopic.

As compared to citric acid, it is slightly less soluble but is still readily soluble in water with a solubility of 132 g per 100 ml at 20°C. It has a stronger apparent acid taste and has a longer taste retention than citric acid which peaks faster but does not mask the aftertaste as effectively. A quantity of 0.362 to 0.408 kg of malic acid is equivalent to 0.453 kg of citric acid and to 0.272 to 0.317 kg of fumaric acid in tartness. At temperatures above 150°C it begins to lose water very slowly to yield fumaric acid. It is used in soft drinks, dry-mix beverages, puddings, jellies, and fruit filling. It is used in hard candies because it has a lower melting point (129°C) than citric acid which improves the ease of incorporation.

Malt A source of the enzyme alpha-amylase which hydrolyzes starch to fermentable sugars such as dextrins and maltose. It is produced by the controlled sprouting of grains, usually barley, followed by drying to produce three general classes of malt differing in amylase content. These classes are brewer's malt, distiller's malt, and gibberellin malt. Malt is used in the brewing industry and as a supplement to flour to increase the alpha-amylase content.

Malted Barley The barley produced under the controlled sprouting of the barley grain followed by drying to obtain the formation of high levels of alpha-amylase and some increase in beta-amylase. These enzymes hydrolyze starch to dextrins and maltose. There are three general classes of malt: brewer's malt, distiller's malt, and gibberellin malt. It is principally used in the brewing industry. In doughs, the malt increases the fermentation rate and improves baking properties.

Malted Cereal Syrup The syrup obtained from barley and other grains, as differentiated from malt syrup which is obtained only from barley. It is used to contribute flavor in baked goods and is a nutrient in yeast fermentation. It is also termed extract of malted barley and corn.

Malt Extract A flavorant formed by extracting the water-soluble enzymes from barley and evaporating to form a concentrate that contains *D*-alpha-amylase enzyme. This enzyme hydrolyzes starch to dextrin and maltose. It is used to provide malt flavor, and in conjunction with spices, seasonings, and flavors.

Malt Flour The flour prepared by the drying and grinding of barley or wheat sprouted under controlled conditions. It can be used as a malt supplement or converted to malt syrups. The malt functions to modify starch during initial stages of baking.

Maltodextrin The product obtained from the partial acid or enzymatic hydrolysis of starch, in the same manner as corn syrup except the conversion process is stopped at an earlier stage. It has a dextrose equivalent of less than 20 and basically is not sweet and is not fermentable. It has fair solubility. It functions as a bodying agent, bulking agent, texturizer, carrier, and crystallization inhibitor. It is used in crackers, puddings, and candies.

Maltol A flavor enhancer used as a synthetic flavoring substance, the function of which is related to ethyl maltol. It occurs naturally in chicory, cocoa, coffee, and cereals. It does not contribute a flavor of its own, but modifies the inherent flavors. As compared to ethyl maltol, it is one-half to one-sixth as effective. It is less soluble, having a solubility of 1 g in 82 ml of water at 25°C. It has a melting range of 160° to 164°C. It is used to enhance the flavor and aroma of fruit, vanilla, and chocolate flavored foods and beverages. It is also used in beverages and desserts with a typical usage range of 10 to 200 parts per million.

Maltose A sweetener formed by the enzymatic action of yeast on starch. It consists of two dextrose molecules. Maltose dissolves and crystallizes slowly in aqueous solutions, and is less sweet and more stable than sucrose. It is used in combination with dextrose in bread and in instant foods, and is also used in pancake syrups.

Malt Syrup The syrup obtained from barley by extraction and evaporation of the worts to 80 to 81 percent solids. It is used as a malt flavor component, as a source of malt and protein, and to provide color. It is used in bakery goods such as rolls and bagels at 1 to 3 percent of the flour weight, in soybean milk at 3 to 7 percent, and in malt base at 1 to 3 percent.

Malt Vinegar A vinegar made by the alcoholic and subsequent acetous fermentation of an infusion of malted barley and/or cereals or a concentrate thereof, which has been enzymatically converted by the malting process. It contains a minimum of 4 percent acid content expressed as acetic acid and is brown to dark brown in color. It functions as an acidulant and preservative in foods.

Manganese A metallic element that functions as a nutrient and dietary supplement. It is necessary for normal bone and tendon structure, central nervous system functionality, and glucose utilization. Sources include manganese carbonate, chloride, oxides, and sulfates. These sources differ in solubility.

Manganese Chloride A source of manganese that functions as a nutrient and dietary supplement. It exists as crystals which are readily soluble in room temperature (22°C) water and are very soluble in hot water. See **Manganese.**

Manganese Citrate A nutrient supplement that is a pale orange or pinkish white powder. It is obtained by precipitating manganese carbonate from manganese sulfate and sodium carbonate solutions. The filtered and washed precipitate is digested first with sufficient citric acid solutions to form manganous citrate and then with sodium citrate to complete the reaction. It is used in baked goods, nonalcoholic beverages, dairy product analogs, fish products, and poultry products. The ingredient may be used in infant formulas.

Manganese Gluconate A nutrient supplement that is a slightly pink-colored powder. It is obtained by reacting manganese carbonate with gluconic acid in aqueous medium and then crystallizing the product. The supplement is used in baked goods, nonalcoholic beverages, dairy product analogs, fish products, meat products, milk products, and poultry products. The ingredient may be used in infant formulas.

Manganese Sulfate A source of manganese that functions as a nutrient and dietary supplement. It exists as a powder which is readily soluble in water. See **Manganese.**

Mannitol A polyol (polyhydric alcohol) that functions as a sweetener, humectant, and bulking agent. It has low hygroscopicity and poor oil solvency. It is approximately 22 percent soluble in water and is approximately 72 percent as sweet as sugar, exhibiting a cool, sweet taste. It functions as a dusting agent with starch in chewing gum. It is used in sugarless candy, chewing gum, cereal, and pressed mints.

Maple Sugar A sweetener obtained by concentrating the sap of the maple sugar tree. It consists of approximately 95 percent sucrose, 2 percent invert sugar, and ash. This is the dry form of maple syrup which is used in syrups and candy.

Maple Syrup A sweetener made by concentrating the sap of the sugar maple tree by boiling at atmospheric pressure. The characteristic color and flavor are developed by heating above 100°C. The concentration at reduced pressure or by freeze-drying gives a sweet, colorless syrup. The characteristic flavor is derived from the volatile oil in the sap. On a dry basis it is approximately 95 percent sucrose, 2 percent invert sugar, and ash. It is used in syrups and candies.

Margarine A product whose consistency and application are similar to those of butter. It is made by emulsifying vegetable oils with milk, followed by crystallization and kneading. Vegetable oils or mixtures of vegetable oils and animal fat may be used. It contains not less than 80 percent fat and is also termed oleomargarine. It is used as a spread and as a source of fat for baked goods and desserts.

Marjoram A spice that is the dried leaves of the herb *Majorana hortensis* Moench. It has a mellow flavor and is distinctively aromatic. The flavor is close to that of oregano. Marjoram is used in soups, sauces, meats, and fish.

Methacrylic Acid-Divinylbenzene Copolymer A carrier of vitamin B_{12} in foods for special dietary use, produced by the polymerization of methacrylic acid and divinylbenzene. The divinylbenzene functions as a cross-linking agent and constitutes a minimum of 4 percent of the polymer.

Methyl B-Naphthyl Ketone (2′-acetonaphtone) A flavoring agent that is a crystalline solid (white or nearly-white), with orange blossom-like odor. It is soluble in most fixed oils, slightly soluble in mineral oil and propylene glycol, and insoluble in glycerin. It is obtained by chemical synthesis.

Methyl Cyclopentenolone (3-Methyl-Cyclopentane-1,2-Dione) A flavoring agent that is a white crystalline powder. It has a nutty odor suggesting a maple-licorice aroma when diluted. It is soluble in alcohol and propylene glycol, slightly soluble in most fixed oils, and sparingly soluble in water. It is obtained by synthesis.

Methylcellulose A gum composed of cellulose in which the methoxyl groups replace the hydroxyl groups. It is soluble in cold water but insoluble in hot water. Solutions increase in viscosity upon heating, gel at 50 to 55°C, and liquefy upon cooling. It is used in baked goods for moisture retention, and in fruit pie fillings for the reduction of water absorption into the pie crust during baking. It is also used in breaded shrimp where it functions to form an oil barrier film.

Methyl Ethyl Cellulose An aerating, emulsifying and foaming agent. The metyoxy content should be not less than 3.5 percent and not more than 6.5 percent, and the ethoxy content should be not less than 14.5 percent and not more than 19 percent, both measured on the dry sample.

Methyl p-Hydroxybenzoate See **Parabens.**

Methyl 2-Methylthiopropionate A synthetic flavoring agent that is a colorless liquid of slightly fruity odor with a suggestion of sulfur. It polymerizes slowly and should be stored in glass or tin containers. It is used in pineapple flavors to give the true note of pineapple. It has applications in beverages, ice cream, candy, and baked goods at 0.5 to 1 part per million.

Methylparaben An antimicrobial agent which is a white free-flowing powder. It is active against yeast and molds over a wide pH range. See **Parabens.**

Methyl Polysilicone See **Dimethylpolysiloxane.**

Methyl Silicone See **Dimethylpolysiloxane.**

3-Methylthiopropionaldehyde A synthetic flavoring agent that is a colorless to light yellow liquid with an intense meat odor. It polymerizes with age and is stable in a 50 percent alcohol solution. It should be stored in glass containers. It is used at low concentrations for meat and broth flavors for applications in meats and condiments at 3 parts per million and in baked goods and beverages at 0.5 part per million.

p-Methoxybenzaldehyde (Anise Aldehyde; p-Anisaldehyde). A flavoring agent that is a colorless or faintly-yellow liquid, hawthorn-like odor. It is miscible in alcohol, ether, and most fixed oils, soluble in propylene glycol, insoluble in glycerin, water, and mineral oil. It is obtained by synthesis.

Microcrystalline Cellulose A gum that is the nonfibrous form of cellulose, an alpha-cellulose. It is dispersible in water but not soluble, requiring considerable energy to disperse and hydrate. In this form it is used in dry applications such as tableting, capsules, and shredded cheese where it functions as a non-nutritive filler, binder, flow aid, and anticaking agent. By addition of carboxymethylcellulose to the alpha-cellulose prior to drying, improved functional properties of hydration and dispersion are obtained. This product is designed for use in water dispersions, being insoluble in water but dispersing in water to form colloidal sols below 1 percent and white opaque gels above the 1 percent usage level. It is used as a heat shock stabilizer and bodying agent in frozen desserts, as an opacifier in low-fat dressings, as a foam stabilizer in whipped toppings, and as an emulsifier in dressings.

Microparticulated Protein Product A fat replacer prepared from egg whites or milk protein or combination egg whites and milk protein. These protein sources may be used alone or in combination with other safe and suitable ingredients to form the microparticulated product. The mixture of ingredients is high-shear heat processed to achieve a smooth and creamy texture similar to that of fat. The ingredient is used in food as a thickener or as a texturizer. It is used in frozen desserts, cheese, dressings, baked goods, and dairy products.

Milk The natural secretion of the mammary glands of female mammals for the feeding of their young. It is commercially considered here as cows' milk which consists, on the average, of 3.5 percent fat, 5 percent lactose, 3.5 percent protein, and 0.7 percent ash. It has a bland, slightly sweet flavor, a yellowish-white color, and a specific gravity of 1.032. It functions as a base for ice cream, yogurt, beverages, and cheese. It is also the source of skim milk, cream, whey, casein, lactose, and milk solids—not-fat.

Milk Albuminate See **Lactalbumin.**

Milk Chocolate See **Chocolate.**

Milkfat The fat of milk which exists in milk as an emulsion of small fat globules in an aqueous phase. It is the only fat in which butyric acid is a component of the glycerides. It has a delicate and pleasant flavor. Approximately 95 percent of the total milk lipids are triglycerides. The average fat content of milk is 3.5 to 3.8 percent. It is used as a source of fat in bakery products, confections, and frozen desserts. It is also termed butter fat.

Milkfat, Anhydrous See **Butter Oil.**

Milk Powder The dry, whole milk that is produced by a spray- or roller-drying process to remove the water fraction. The dry form offers convenience of transportation, utility, and stability. It is used in soup mixes and dessert mixes.

Milk Solids—Not-Fat The dry form of skim milk. It contains not more than 1.5 percent fat and not more than 5 percent moisture. It has excellent flavor, nutritional value, and functional properties such as water binding, emulsification, and foam formation. It is used in ice cream mix, baked goods, and desserts. It is also termed nonfat dry milk, skim milk powder, and dried skim milk.

Milk Sugar See **Lactose.**

Mint A spice derived from any one of the plants of the mint family (Labiatae) of which there are numerous varieties. Only spearmint and peppermint are commercially important. Mint is used in mint sauce, fruit cocktails, and beverages in either its dried or fresh forms.

Modified Food Starch See **Modified Starch.**

Modified hop extract A flavoring agent in the brewing of beer. It is manufactured from a hexane extract of hops, with simultaneous isomerization and selective reduction in an alkaline aqueous medium with sodium borohydride. It is added to the wort before or during cooking in the manufacture of beer.

Modified Starch The product resulting from the treatment of starch with certain chemicals to modify the physical characteristics of the native starch. This produces more desirable or useful characteristics such as improved solubility, acid stability, and texture. It is used in desserts, pie fillings, sauces, gravies, and fabricated foods as a thickener, binder, and stabilizer.

Molasses The by-product of the manufacture of sugar from sugar cane in which the syrup is separated from the crystals. The highest grade is edible molasses which is most often found as table syrup. The lowest grade is blackstrap molasses. Molasses is a strongly flavored, dark colored syrup containing 70 to 80 percent solids of which 50 to 75 percent is sucrose and invert sugar. It is used in syrups and in the production of caramel.

Monoammonium L-Glutamate (monoammonium glutamate; ammonium glutamate). A flavor enhancer and salt substitute that is crystalline powder (white, free-flowing), and odorless. It is soluble in water, insoluble in common organic solvents, and is obtained by chemical synthesis.

Monoammonium Glycyrrhizinate See **Glycyrrhizin.**

Mono- and Diglycerides A lipophilic emulsifier that consists of both monoglycerides and diglycerides. It is made by reacting glycerol with specific fats or oils. The consistency varies from yellow liquid to ivory-colored plastic to hard solids of bland odor and taste. It is the most commonly used emulsifier in foods. It is used in numerous applications for example in baked goods, frozen desserts, whipped toppings, and margarine for a variety of functions. Typical usage levels range from 0.05 to 0.50 percent.

Monocalcium Phosphate An acidulant, leavening agent, and nutritional supplement that exists as white crystals or granular powder. It is sparingly soluble in water. It is used as an acidulant in breads and dry-mix beverages; as a source of calcium in fruit jellies, preserves, and cereals; and as a component of dough conditioners. It is also of restricted use as a chemical leavening agent because it releases about 67 percent of the carbon dioxide during the initial mixing and this is generally too rapid. It is also termed calcium phosphate monobasic, calcium biphosphate, and acid calcium phosphate.

Monoglyceride A lipophilic emulsifier prepared by the direct esterification of fatty acids with glycerol or by the interesterification between glycerol and other triglycerides. It is insoluble in water. It provides emulsion stability, prevents fat separation, and also functions as a foaming agent, defoaming agent, and dispersant. It is most often used in combination with a diglyceride. It is used in ice cream, peanut butter, puddings, and numerous other applications. The typical usage level is 0.05 to 0.40 percent.

Monoglyceride Citrate A sequestrant that is a mixture of glyceryl monooleate and its citric acid mono-ester. It is soluble in common fat solvents and alcohol and is insoluble in water. It is used in antioxidant formulations for addition to fats and oils at less than 200 parts per million. It functions as a antioxidant synergist in peanut oil at a maximum usage level of 100 mg/kg individually or in combination. In margarine, it is used at no more than 0.01 percent.

Monoglyceride, Distilled See **Distilled Monoglyceride.**

Monoisopropyl Citrate A sequestrant used in fats and oils.

Monopotassium Monophosphate See **Monopotassium Phosphate.**

Monopotassium Phosphate A buffer, neutralizing agent, and sequestrant. It is mildly acid, with a pH of 4.5, and fairly soluble in water, with a solubility of 25 g in 100 ml at 25°C. It is used in whole eggs for color preservation and is also used in low-sodium products, milk products, and meat products. Typical usage ranges from 0.1 to 0.5 percent. It is also termed potassium dihydrogen orthophosphate, potassium phosphate monobasic, and monopotassium monophosphate.

Monosodium Dihydrogen Orthophosphate See **Monosodium Phosphate.**

Monosodium Glutamate (MSG) A flavor enhancer that is the sodium salt of glutamic acid, an amino acid. It is a white crystal that is readily soluble in water. It intensifies and enhances flavor but does not contribute a flavor of its own. It may be present as one of the amino acids or in a free form, which is how it effectively enhances the flavor of foods. It is produced through a fermentation process of molasses. It is used at 0.1 to 1.0 percent in meats, soups, and sauces.

Monosodium Monophosphate See **Monosodium Phosphate.**

Monosodium Phosphate An acidulant, buffer, and sequestrant that is mildly acid, with a pH of 4.5, and very soluble in water, with a solubility of 87 g per 100 ml of water at 25°C. It is used as an acidulant in effervescent powders and laxatives. It is also used in soft drink dry-mix formulations, in cheese, and in carbonated beverages. It is also termed monosodium dihydrogen orthophosphate; sodium phosphate, monobasic; sodium biphosphate; and monosodium monophosphate.

Monosodium Phosphate Derivatives of Mono- and Diglycerides See **Mono- and Diglycerides.**

Mono-Tertiary-Butylquinone See **Tertiary Butylhydroquinone.**

Mustard A flavorant made from the dried, ripe seed of several closely related genera, species, and varieties of the family Cruciferae; the seed of a plant of the cabbage family. It is used as a flavorant in baked goods, sauces, salad dressings. It also functions as an emulsifier in salad dressings. The ground seed is used for spices.

Mustard Flour The ground seed of the mustard plant from which some of the oil and most of the hulls have been removed. It is used in salad dressings and sauces, and as a condiment.

Mustard Oil See **Allyl Isothiocyanate.**

Mustard Seed A spice of which there are several varieties, the dry mustards being of the hot or mild type. It is used in meats, sauces, and salad dressings.

Myristic Acid A fatty acid obtained from coconut oil and other fats. It has poor water solubility but is soluble in alcohol, chloroform, and ether. It is used as a lubricant, binder, and defoaming agent.

N

Natamycin A preservative for use as a coating on the surface of Italian cheeses to prevent the growth of mold or yeast. It is tasteless, odorless, colorless, and does not penetrate the cheese. It is very active against virtually all molds and yeasts, but does not affect bacteria, thus not affecting the ripening and flavor improvement process of cheese. it can be applied as a dip, spray, or by other methods such as incorporation into the cheese coatings. It is used at levels ranging from 300 to 2000 parts per million. It is also termed pimaricin.

Natural Sugar See **Turbinado Sugar.**

Nerol (Cis-3—,7-Dimethyl-2,6-Octadien-1-OI) A flavoring agent that is a colorless liquid with an odor resembling fresh, sweet roses and contains geranoils and other terpenic alcohols. It is miscible in alcohol, chloroform, and ether; insoluble in water. It is obtained by synthesis.

Niacin A water-soluble B-complex vitamin that is necessary for the growth and health of tissues. It prevents pellagra. It has a solubility of 1 g in 60 ml of water and is readily soluble in boiling water. It is relatively stable in storage and no loss occurs in ordinary cooking. Sources include liver, peas, and fish. It was originally termed nicotinic acid and also functions as a nutrient and dietary supplement.

Niacinamide A nutrient and dietary supplement that is an available form of niacin. Nicotinic acid is pyridine beta-carboxylic acid and nicotinamide, which is another term for niacinamide, is the corresponding amide. It is a powder of good water solubility, having a solubility of 1 g in 1 ml of water. Unlike niacin, it has a bitter taste.

Nicotinamide See **Niacinamide.**

Nicotinic Acid See **Niacin.**

Nisin An antimicrobial agent derived from pure culture fermentations of certain strains of *Streptococcus lactis* Lancefield Group N. Nisin preparation contains nisin, a group of related peptides with antibiotic activity. It is used to inhibit the outgrowth of *Clostridium botulinum* spores and toxin formation in pasteurized cheese spreads and pasteurized process cheese spreads; pasteurized cheese spread with fruits, vegetables, or meats; and pasteurized process cheese spread with fruits, vegetables, or meats.

Nitrate The salt of nitric acid. It is used in meat curing to develop and stabilize the pink color associated with cured meat. By itself, it is not effective in producing the curing reaction until it is chemically reduced to nitrite. It has an effect on flavor and also functions as an antioxidant. It is available as sodium and potassium nitrate, with the sodium form being more common.

Nitrite The salt of nitrous acid. It is used in meat curing to develop and stabilize the pink color associated with a cured meat and to affect flavor and function as an antioxidant. Nitrites convert to nitric oxide, which reacts with the myoglobin pigments (purple-red) to form nitro-somyoglobin (dark red). Nitrosomyoglobin plus heating to 130° to 140°F results in the formation of the stable pigment nitrosohemochrome, resulting in the cured meat color. It has bacteriostatic properties as an inhibitor of *Clostridia* organisms, especially *Clostridium botulinum,* and, therefore, nonsterile canned hams can be produced. Sources are sodium and potassium nitrite, with the sodium form being more commonly used.

Nitrous Oxide A noncombustible gas used as a propellant in certain dairy and vegetable fat whipped toppings contained in pressurized containers.

(Gamma)-Nonalactone A synthetic flavoring agent that is a colorless to yellow liquid of strong, coconut-like odor. It is soluble in most fixed oils, mineral oil, and propylene glycol. It is stable in acids and unstable in alkali and should be stored in glass, tin, or aluminum containers. It is used in coconut flavors and has application in gelatins, puddings, baked goods, candy, and ice cream at 11 to 55 parts per million. It is also termed aldehyde C-18.

Nonanal (Aldehyde C-9; Pelargonic Aldehyde) A flavoring agent that is a colorless or light-yellow liquid, with a strong odor resembling an essence of orange and rose. It is soluble in alcohol, most fixed oils, mineral oil, and propylene glycol, but insoluble in glycerin. It is obtained by chemical synthesis.

Nonfat Dry Milk See **Milk Solids—Not-Fat.**

Norbixin See **Annatto.**

Nordihydroguaiaretic Acid (NDGA) An antioxidant that has poor solubility and shows evidence of discoloration in the presence of metal salts. It is used to a limited extent to retard rancidity.

Nutmeg A spice obtained from the nutmeg tree *Myristica fragrans.* It is related to mace, which is obtained from the covering of nutmeg. Nutmeg is used in eggnog, cakes, fruit, and puddings.

O

Oat A grain that is a source of oat flour. It is used in porridge, grits, and oatmeal.

Oat Flour Fine-mesh ground oats with the hull removed. It has some antioxidant properties and is blended with other flours to retard rancidity.

Oatmeal The food produced by grinding oats after removal of the husk.

(Gamma)-Octalactone A synthetic flavoring agent that is a stable, colorless to slightly yellow liquid of peach odor. It should be stored in glass or tin containers. It is used in flavors for peach with applications in baked goods, candy, and ice cream at 5 to 17 parts per million.

1-Octanol A synthetic flavoring agent that is a colorless, stable liquid of sharp fatty odor. It is soluble in alcohol, most fixed oils, mineral oil, and propylene glycol. It should be stored in glass or tin containers. It is used in essential oils for application in beverages, candy, and baked goods at 1 to 3 parts per million. It is also termed octyl alcohol.

1-Octanol, Natural (Alcohol C-8; Octyl Alcohol; Capryl Alcohol) A flavoring agent that is a colorless liquid, with a penetrating fat-like odor. It is soluble in most fixed oils, mineral oil, and propylene glycol, but insoluble in glycerin. It is obtained from natural precursors.

Octyl Acetate A flavoring agent that is a colorless liquid with a fruity odor resembling orange and jasmine. It is miscible in alcohol, oils, and other organic solvents, and insoluble in water. It is obtained by chemical synthesis.

Octyl Alcohol See **1-Octanol.**

Oil of Rue A flavoring agent that is the natural substance obtained by steam distillation of the fresh blossoming plants of rue, the perennial herb of several species of *Ruta (Ruta montana* L., *Ruta graveolens* L., *Ruta bracteosa* L., and *Ruta calepensis* L.). It is used in baked foods and baking mixes (10 ppm); frozen dairy desserts and mixes (10 ppm); soft candy (10 ppm); and other food categories (4 ppm).

Oleic Acid An unsaturated fatty acid that functions as a lubricant, binder, and defoamer.

Oleic Acid Derived from Tall Oil Fatty Acids An additive consisting of purified oleic acid separated from refined tall oil fatty acids. It is used in foods as a lubricant, binder, and defoaming agent, and as a component in the manufacture of other food-grade additives. To assure safe use of the additive, the label should show the common or usual name of the acid, and the words "food grade."

Oleomargarine See **Margarine.**

Oleoresins Solvent-free extractions from spices that contain the volatile and nonvolatile flavor components. They are the closest replacements for a spice, and are used in seasonings for a variety of foods.

Oleoresin Paprika A seasoning and colorant that is the solvent-free extraction containing the volatile and nonvolatile flavor components of paprika. It is the closest replacement for paprika. As a colorant, the pigment is a red-orange carotenoid of which the principal carotenoid is capsanthin. It has fair pH and heat stability, and poor light and chemical stability. It is used in sausages, meat products, condiment mixtures, and salad dressings.

Olive Oil The oil obtained from the fruit of olive trees, *Olea europaea*. It is used mainly for salad and cooking oils.

Onion A flavorant, the vegetable *Allium cepa* L., commercially processed into powder, salt, minced, and toasted forms. It is used in meats, sauces, soups, and dips.

Orange Oil, Bitter A flavoring agent that is a yellow-brown liquid with an aromatic odor resembling Seville orange, and an aromatic and bitter taste. Its substance is degraded by light, and its alcohol solutions are neutral to litmus. It is miscible in absolute alcohol and glacial acetic acid, soluble in fixed oils and mineral oil, slightly soluble in propylene glycol, and insoluble in glycerin. It is obtained by cold expression of fresh peel of the fruit of *Citrus aurantium* L. of the *Rutaceae* family.

Oregano A spice made from the dried leaves of *Lippia graveolens,* a perennial of the mint family. There are two strains available. One strain, common to the Mediterranean region, is delicate in fragrance and taste and the other, which is common to Mexico, is quite pungent. It is used in sauces, soups, and pizza.

Orthophosphoric Acid See **Phosphoric Acid.**

Ox Bile Extract A yellowish green, soft solid, with a part-sweet, part-bitter, disagreeable taste. It is the purified portion of the bile of an ox obtained by evaporating the alcohol extract of concentrated bile. The ingredient is used as a surfactant in food, a surfactant also known as purified oxgall or sodium choleate.

Oxidized Corn Starch Starch produced by treating an aqueous starch suspension with dilute sodium hypochlorite containing a small excess of caustic soda until the desired degree of oxidation is reached. The slurry is then treated with an antichlor, such as sodium bisulfate, adjusted to the desired pH, filtered, washed, and dried. It still retains its original granule structure and is insoluble in water. It is extremely white, has decreased viscosity, is relatively clear, and shows a reduced tendency to thicken when cooled. Its food applications are those where high solids and low viscosity are desired.

Oxidizing and Reducing Agents Substances which chemically oxidize or reduce another food ingredient, thereby producing a more stable product.

Oxystearin A crystallization inhibitor and release agent that is a modified fatty acid composed of the glycerides of partially oxidized stearic and other fatty acids. It is used in vegetable oils to prevent them from clouding in the refrigerator and in griddling fats and oils to prevent food from sticking to the frying pan.

P

P-Methoxybenzaldehyde (Anisic Aldehyde; p-Anisaldehyde). A flavoring agent that is a colorless or faintly-yellow liquid, hawthorn-like odor. It is miscible in alcohol, ether, and most fixed oils, soluble in propylene glycol, insoluble in glycerin, water, and mineral oil. It is obtained by synthesis.

Palmitic Acid A fatty acid which is a mixture of solid organic acids from fats consisting principally of palmitic acid with varying amounts of stearic acid. It functions as a lubricant, binder, and defoaming agent.

Palm Kernel Oil An oil obtained from palm kernels. It consists mainly of lauric, myristic, and oleic fatty acids. It resembles coconut oil and is used interchangeably with coconut oil. It is a possible source of stearine, which is a substitute for cocoa butter. It is used in margarine and confectionary.

Palm Oil The oil obtained from the fruit of the palm tree. It has a narrower plastic range than lard and most shortenings which is a disadvantage in shortening applications. It can be used in mixtures with only a moderately adverse effect on the plastic range. It consists mainly of palmitic, oleic, and linoleic fatty acids. It is used in margarine and shortenings.

Pantothenic Acid Vitamin B_5, which is a water-soluble vitamin. It is required for proper growth and maintenance of the body and is involved in body processes such as energy release from carbohydrates and metabolism of fatty acids. It is relatively stable through storage and is found in liver, eggs, and meat.

Papain A tenderizer that is a protein-digesting enzyme obtained from the papaya fruit. The enzyme, used in a patented process, is injected into the circulatory system of the live animal and is activated by the heat of cooking to break down the protein, thus tenderizing the beef. The enzyme is inactivated by stomach acids.

Paprika A spice and colorant made from the ground, dried, ripe fruit of the herb *Capsicum annuum* L. It contributes flavor and color to foods. The pod provides red color and has good tinctorial strength, good pH stability, and poor stability to light and oxidation. It is used in meat, fish, sauces, and salad dressings. It is also termed sweet pepper or pimiento. Also see Oleoresin Paprika.

Parabens Antimicrobial agents that are esters of para-hydroxybenzoic acid. The most common esters are methyl p-hydroxybenzoate and propyl p-hydroxybenzoate. The ethyl and butyl esters have some applications. It is related to benzoic acid and sodium benzoate but is effective over a wide pH range. The parabens are most active against yeasts and molds and are stable to high temperature. They are a white free-flowing powder of fair water solubility at room temperature which improves if the water is heated to 70°C. Methyl paraben is more soluble (0.25 g per 100 ml of water at 25°C) but less effective in mold inhibition than propyl paraben (0.04 g per 100 ml of water at 25°C). It is used in meat and poultry products.

Para-Hydroxybenzoic Acid See **Parabens.**

Parboiled Rice The rice that results from the process of soaking rice in water, draining, pressure cooking to completely gelatinize the starch, drying, and milling. The parboiling process aids the development of stability toward cooking and heat processing. It is used in canned rice products such as soups, casseroles, meat, and rice dinners, such as Spanish rice. The milling of parboiled rice produces parboiled bran.

Parsley A spice made from the dried leaves of *Petroselinum hortense,* of bright green color. It has a high content of vitamins A and C and also contains iron, iodine, copper, and manganese. It is used for garnishing and seasoning, with application in sauces, salads, and soups.

Partially Hydrogenated Coconut Oil See **Coconut Oil.**

Partially Hydrogenated Oil See **Hydrogenated Vegetable Oil.**

Pastry Flour A flour obtained from soft wheat. Either straight or clear flour grades may be used because color is not an essential requirement. It is used in white sauces and pastry.

Patent Flour Flour made from the separation of 40 to 90 percent of that portion of the grain that can be milled from a wheat blend. There are various streams to include long patent, medium patent, short patent, first patent, and fancy patent flours.

Peanut Oil The oil obtained from peanuts, consisting principally of the unsaturated fatty acids oleic and linoleic. It is liquid at room temperature, has a specific gravity at 38°C of approximately 1.89 to 0.90, and an iodine number of 85 to 95. It is removed from the nuts by one of two processes, namely, the expeller method, in which the shelled peanuts are cooked with steam, and fed into an expeller press which physically presses the oil from the meal; or the pre-press solvent system, which is comparable to the expeller method except that less pressure is applied, which leaves more oil in the meal, and the remaining meal is solvent-washed, usually with hexane, to dissolve the oil from the meal. The obtained crude oil is refined. The major use of peanut oil is in cooking oils and salad oils. Peanut oil is used in deep-fat frying because of its long frying life and high smoke point. In salad oil, it contributes to the suspension of solids. Other applications include shortening ingredient for doughnuts and cakes.

Pearl Starch See **Cornstarch.**

Pectic Acid Those pectic substances that are essentially void of methoxyl groups and have carboxyl groups only. They have varying degrees of neutralization. The divalent salts are slightly soluble in water and must be converted to the sodium or potassium forms for dissolution. It gels in the presence of calcium or other divalent cations.

Pectin A gum that is water-soluble pectinic acid of varying methyl ester content and degree of neutralization. It is obtained from citrus peel and apple pomace. It forms a gel in systems of low pH (pH 2.8–3.7) and high sugar (55–80 percent) levels. The gel sets at 55° to 99°C and melts above 70°C. Pectins are characterized by rapid and slow set types. The high-methoxyl pectins have a degree of methylation (DM) greater than 50 percent, while those of less than 50 percent degree of methylation are termed low-methoxyl pectins. The low-methoxyl pectins gel in the presence of calcium ions and do not require a certain level of acid or sugar.

It is used in beverages at 0.1 to 0.2 percent, in jams and jellies at 0.1 to 0.4 percent, and in confectioner's jelly at 0.8 to 1.5 percent. See also amidated pectin.

Pectinic Acid A broad group of pectic substances that contain more than a negligible proportion of methyl ester groups and all the unesterified carboxyl groups are free. The divalent salts of pectinic acid are only slightly soluble in water and must be converted to the sodium or potassium form for dissolution. See **Pectin.**

2-Pentanone (Methyl Propyle Ketone) A flavoring agent that is a clear liquid, colorless, with flowery odor. It is miscible in alcohol and ether and soluble in water. It is obtained by chemical synthesis.

Pentasodium Tripolyphosphate See **Sodium Tripolyphosphate.**

Pepper A spice made from a berry from the vine *Piper nigrum* L. which produces black and white pepper. Black pepper is picked slightly underripe and dried, during which time the characteristic black, wrinkled appearance is attained. White pepper is picked fully ripe and dried, after which the outer hull is removed by attrition to expose the white core. It is used in meat, vegetables, soups, and salads.

Pepper, Cayenne A spice that is not related to the true pepper vine but to the paprika, bell peppers of the Capsicum family. It is hot and fiery and used in spreads, dips, and chili sauce.

Pepper, Red The pod of the genus *Capsicum,* variety *C. annuum* L. and *C. frutescens* L. It has a hot, pungent flavor. It is used in barbecue sauce, spicy sauces, and chili powder.

Peptone A polypeptide used as a beer stabilizer.

Petrolatum A release agent, lubricant, and defoaming agent that is a purified mixture of semisolid hydrocarbons obtained from petroleum. It varies in color from white to yellow. It is used in bakery products, dehydrated fruits and vegetables, and egg white solids.

Petroleum Wax A wax used as a masticatory substance in chewing gum base. It is also used as a protective coating on raw fruits and vegetables and as a fruit defoamer.

Phenethyl Phenylacetate A flavoring agent that is a colorless or pale yellow liquid, with an odor resembling roses and hyacinth, which becomes solid at <26°C (78.8°F). It is soluble in alcohol, insoluble in water. It is obtained by chemical synthesis.

Phenylacetic Acid (A-Toluic Acid) A flavoring agent that is crystalline (white, glistening), with unpleasant, persisting odor resembling geranium leaf and rose when diluted. It is soluble in most fixed oils and glycerin, slightly soluble in water, and insoluble in mineral oil. It is obtained by chemical synthesis.

Phenylethyl Anthranilate A synthetic flavoring agent that is a stable, white to yellow crystal of grape and orange blossom odor. It should be stored in glass or polyethylene-lined containers. It is used for flavors such as grape and cherry in applications such as beverages, ice cream, candy, and baked goods at 2 to 6 parts per million.

Phenylethyl Isobutyrate A synthetic flavoring agent that is a stable, colorless to light yellow liquid of fruity odor of floral note. It is soluble in alcohol and practically insoluble in water. It should be stored in glass or tin-lined containers. It is used in flavors for peach with applications in beverages, ice cream, candy, and baked goods at 3 to 13 parts per million.

Phosphate Any salt of phosphoric acid. The salts include disodium phosphate, trisodium phosphate, sodium hexamethaphosphate, and others. They play a variety of roles such as sequestrants, emulsifiers, solubility enhancers, and buffers in a variety of foods.

Phosphated Flour Flour to which monocalcium phosphate is added at not less than 0.25 percent and not more than 0.75 percent. It is used in baked goods.

Phosphoric Acid An acidulant that is an inorganic acid produced by burning phosphorus in an excess of air, producing phosphorus pentoxide which is dissolved in water to form orthophosphoric acid of varying concentrations. It is a strong acid which is soluble in water. The acid salts are termed phosphates. It is used as a flavoring acid in cola and root beer beverages to provide desirable acidity and sourness. It is used as a synergistic antioxidant in vegetable shortenings. In yeast manufacture, it is used to maintain the acidic pH and provide a source for phosphorus. It also functions as an acidulant in cheese. It is also termed orthophosphoric acid.

Pimaricin See **Natamycin.**

Pimiento See **Paprika.**

Piperonyl Acetate A synthetic flavoring agent that is a stable, colorless to light yellow liquid of heliotrope odor. It should be stored in glass or resin-lined containers. It is used in flavors for berry notes with applications in beverages, candy, ice cream, and baked goods at 50 to 90 parts per million.

Piperonyl Isobutyrate A synthetic flavoring agent that is a moderately stable, colorless to light yellow liquid of fruity odor. It should be stored in glass or resin-lined containers. It is used in flavors for cherry, berry, and peach aroma with applications in beverages, candy, and baked goods at 1 to 4 parts per million.

Plasticizer See **Softener.**

Polydextrose A bulking agent that is a randomly bonded condensation polymer of dextrose containing small amounts of bound sorbitol and citric acid. It is a water-soluble powder providing a pH range of 2.5 to 3.5. It is partially metabolized which results in a caloric value of one calorie per gram. As a reduced-calorie bulking agent, it can partially replace sugars and in some cases fats in reduced-calorie foods. It also functions as a bodying agent and humectant. Applications include desserts, specific baked goods, frozen dairy desserts, chewing gum, and candy. Usage levels vary according to application, but examples are frozen dessert, 13 to 14 percent; puddings, 8 to 9 percent; and cake, 15 to 16 percent.

Polyethylene Glycol A binder, coating agent, dispersing agent, flavoring adjuvant, and plasticizing agent that is a clear, colorless, viscous, hygroscopic liquid resembling paraffin (white, waxy, or flakes), pH = 4.0–7.5 (1:20 concentration). It is soluble in water (MW 1000) and many organic solvents.

Polyglycerate 60 See **Ethoxylated Mono- and Diglycerides.**

Polyglycerol Esters of Fatty Acids Emulsifiers that are mixed partial esters formed by reacting polymerized glycerols with edible fats, oils, or fatty acids. They vary in degree of polymerization, and by varying the proportions and fats to be reacted, a diverse class of products is obtainable. The esters range from hydrophilic to lipophilic. They are used in cake mixes for volume and texture, in confectionary for gloss, in whipped toppings for aeration, and in flavors and colors as a solubilizer. Typical usage range is from 0.1 to 1.0 percent.

Polyoxyethylene (20) Mono- and Diglycerides of Fatty Acids See **Ethoxylated Mono- and Diglycerides.**

Polyoxyethylene Sorbitan Ester See **Polyoxyethylene Sorbitan Fatty Acid Esters.**

Polyoxyethylene Sorbitan Fatty Acid Esters Emulsifiers made by reacting ethylene oxide with sorbitan esters to increase their hydrophilic properties. They are generally used in oil and water emulsions in combination with lipophilic emulsifiers such as mono- and diglycerides or sorbitan monostearates to produce a wide variety of effects. They are also termed polysorbates, which include polysorbate 80 (polyoxyethylene [20] sorbitan monooleate), polysorbate 60 (polyoxyethylene [20] sorbitan monostearate), and polysorbate 65 (polyoxyethylene [20] sorbitan tristearate). They can solubilize essential and vitamin oils. They are used in panned coatings to reduce panning time, in coffee whiteners to prevent oiling-off, and in ice cream to produce dryness and overrun. Typical usage level ranges from 0.05 to 0.10 percent.

Polyoxyethylene (20) Sorbitan Monooleate An emulsifier produced by reacting oleic acid with sorbitol to yield a product which is reacted with ethylene oxide. It is a nonionic, water-dispersible surface-active agent that is very soluble in water. It is also termed polysorbate 80. It is used in ice cream and frozen desserts for overrun and dryness; as a disperser of flavors and colors in pickles; and for volume and texture in baked goods. It is frequently used with mono- and diglycerides at usage levels ranging from 0.05 to 0.10 percent.

Polyoxyethylene (20) Sorbitan Monostearate An emulsifier manufactured by reacting stearic acid with sorbitol to yield a product which is reacted with ethylene oxide. It is a nonionic, water-dispersible surface-active agent which is very hydrophilic. It is also termed polysorbate 60. It is used in whipped vegetable toppings for overrun and lightness; in cakes for increased volume and fine grain; in icings and confectionary for lightness and syneresis control; and in salad dressing for emulsion stability. It is frequently used with sorbitan monostearate or mono- and diglycerides. The typical usage range is 0.10 to 0.40 percent.

Polyoxyethylene (20) Sorbitan Tristearate An emulsifier manufactured by reacting stearic acid with sorbitol to yield a product which is then reacted with ethylene oxide. It is a nonionic surface-active agent which is dispersible in fat, oil, and water. It is also termed polysorbate 65. It is used in frozen desserts, cakes, and coffee whiteners. It is frequently used with sorbitan monostearates or mono- and diglycerides. Typical usage range is 0.10 to 0.40 percent.

Polyoxyl (40) Stearate An emulsifier and antifoaming agent used in processed foods, fruit jellies, and sauces.

Polysorbates See **Polyoxyethylene Sorbitan Fatty Acid Esters.**

Polysorbate 60 See **Polyoxyethylene (20) Sorbitan Monostearate.**

Polysorbate 65 See **Polyoxyethylene (20) Sorbitan Tristearate.**

Polysorbate 80 See **Polyoxyethylene (20) Sorbitan Monooleate.**

Pomace Ground apple or fleshy fruit in the dry form.

Popcorn Indian corn that explodes when exposed to dry heat due to the expansion of the kernel.

Poppy Seed A seasoning that is a seed of *Papaver somniferum* L. Poppy seeds have a nutty flavor. They are used in breads, cakes, and butter sauce for vegetables, lending a nutlike flavor.

Potassium Acid Tartrate See **Cream of Tartar.**

Potassium Alginate A gum that is the potassium salt of alginic acid. It is soluble in cold water, forming a viscous colloidal solution. It functions as a stabilizer, thickener, and gelling agent. It is used in dietetic foods, low-sodium foods, dry mixes, and dental impression material. Typical usage levels range from 0.05 to 0.50 percent.

Potassium Bicarbonate An alkali and leavening agent obtained as colorless prisms or white powder. It is very soluble, with 1 g dissolving in 2.8 ml of water. Upon heating it liberates carbon dioxide which provides leavening in baked goods. It is also used in confectionary products.

Potassium Bisulfite A preservative that retards bacterial action, prevents discoloration, and functions as an antioxidant. It is not used in meats or in food recognized as a source of vitamin B_1, and it is not used on fruits or vegetables intended to be served raw or presented as fresh.

Potassium Bitartrate See **Cream of Tartar.**

Potassium Bromate A dough conditioner that exists as white crystals or powder and is soluble in water. It exists in the anhydrous form as white granular powder and in the hydrated form as small white crystals or granules. It is used to age and improve the baking properties of flour. It is used with potassium iodate and azodicarbonamide to modify the protein in bread flour to promote the desired properties of loaf volume and shape. It is used in baked goods.

Potassium Carbonate A general purpose food additive and alkali. It is hygroscopic and the aqueous solutions are strongly alkaline. It has a solubility of 1 g in 1 ml of water at 25°C. It is used as a flavoring agent and processing aid, and to control pH. It is used in soups to neutralize acidity.

Potassium Carrageenan See **Carrageenan.**

Potassium Caseinate See **Caseinates.**

Potassium Chloride A nutrient, dietary supplement, and gelling agent that exists as crystals or powder. It has a solubility of 1 g in 2.8 ml of water at 25°C and 1 g in 1.8 ml boiling water. Hydrochloric acid, and sodium chloride and magnesium chloride diminish its solubility in water. It is used as a salt substitute and mineral supplement. It has optional use in artificially sweetened jelly and preserves. It is used as a potassium source for certain types of carrageenan gels. It is used to replace sodium chloride in low-sodium foods.

Potassium Citrate, Monohydrate A sequestrant and buffer that exists as crystals or powder. It is slightly hygroscopic and possesses the advantageous properties of citric acid without having its acid reaction. A 1 percent solution has a pH of 7.5 to 9.0. It reacts with metal ions such as calcium, magnesium, and iron to form a complex. It is soluble in water with a solubility of 1.8 g in 1 ml 20°C water and 2 g in 1 ml of 80°C water. It is found in artificially sweetened jelly and in certain milk and meat products. Uses include processed cheese, puddings, and dietetic foods in which sodium is undesirable. It is also termed tripotassium citrate.

Potassium Dihydrogen Orthophosphate See **Monopotassium Phosphate.**

Potassium Hydrogen Tartrate The acid potassium salt of tartaric acid, which has relatively poor water solubility. It functions to complex with heavy metal ions and regulates pH. See **Cream of Tartar.**

Potassium Hydroxide A water-soluble food additive and bleaching agent. Upon exposure to air it readily absorbs carbon dioxide and moisture and deliquesces. It is used to destroy the bitter chemical constituents in olives that will be used as black olives.

Potassium Iodate A source of iodine made by reacting iodine with potassium hydroxide. It is a crystalline powder which is more stable than iodide. It has a solubility of 1 g in 15 ml of water. It is used as a fast-acting dough improver; it is used with potassium bromate as an oxidizing agent to modify the protein in bread flour which promotes loaf volume and shape. It is used in baked goods.

Potassium Iodide A source of iodine and a nutrient and dietary supplement. It exists as crystals or powder and has a solubility of 1 g in 0.7 ml of water at 25°C. It is included in table salt for the prevention of goiter.

Potassium Lactate A flavor enhancer that is the potassium salt of lactic acid. It is a hydroscopic, white, odorless solid and is prepared commercially by the neutralization of lactic acid with potassium hydroxide. It is also used as a flavoring agent or adjuvant, a humectant, and a pH control agent.

Potassium Metabisulfite A chemical preservative and antioxidant obtained as white or colorless crystals, powder, or granules. It is soluble in water and insoluble in alcohol. The sulfite salt yields sulfurous acid at a low pH. It is used as a food preservative.

Potassium Metaphosphate A substitute for sodium phosphate in low-sodium foods. Also functions as a fermentation nutrient and buffer.

Potassium Nitrate A preservative and color fixative in meats which exists as colorless prisms or white granules or powder. It has a solubility of 1 g in 3 ml of water at 25°C. See **Nitrate.**

Potassium Nitrite A color fixative in meats which exists as white or yellowish granules or cylindrical sticks. It is very soluble in water. See **Nitrite.**

Potassium Oleate The potassium salt of oleic acid. It is used as a binder, emulsifier, and anticaking agent.

Potassium Palmitate The potassium salt of palmitic acid. It is used as a binder, emulsifier, and anticaking agent.

Potassium Phosphate Dibasic See **Dipotassium Phosphate.**

Potassium Phosphate Monobasic See **Monopotassium Phosphate.**

Potassium Sodium Tartrate See **Sodium Potassium Tartrate.**

Potassium Sorbate A preservative that is the potassium salt of sorbic acid. It is a white crystalline powder which is very soluble in water, with a solubility of 139 g in 100 ml at 20°C. This solubility allows for solutions of high concentration which can be used for dipping and spraying. It is effective up to pH 6.5. It has approximately 74 percent of the activity of sorbic acid, therefore requiring higher concentrations to obtain comparable results as sorbic acid. It is effective against yeasts and molds and is used in cheese, bread, beverages, margarine, and dry sausage. Typical usage levels are 0.025 to 0.10 percent.

Potassium Stearate The potassium salt of stearic acid. It is used as a binder, emulsifier, anticaking agent, and as a placticizer in chewing gum base.

Potassium Sulfate A flavoring agent that occurs naturally, consisting of colorless or white crystals or crystalline powder having a bitter, saline taste. It is prepared by the neutralization of sulfuric acid with potassium hydroxide or potassium carbonate.

Potato Starch A starch obtained from potatoes. It provides long body and clarity to food. It is used mainly in those countries in which it is the principal commercial starch. Applications include Danish desserts, soups, and gravies.

Powdered Sugar A sweetener obtained by pulverizing granulated sugar and adding approximately 3 percent cornstarch. The blend is ground to the desired fineness, that is, 4X, 6X, or 8X. It is very soluble in water. Applications include confectionaries and icings.

Precipitated Calcium Phosphate See **Tricalcium Phosphate.**

Pregelatinized Starch Starch that has been processed to permit swelling in cold water, unlike natural starch which requires heating. The processing usually consists of cooking starch slurries, drying, and grinding to a fine powder. It is used in instant puddings, cake mixes, and soup mixes at 1 to 5 percent. It is also termed gelatinized wheat starch.

Preservatives Antimicrobial agents used to preserve food by preventing growth of microorganisms and subsequent spoilage, including fungicides, mold and rope inhibitors. The preservatives most widely used are the benzoates (sodium benzoate), sorbates (sorbic acid and potassium sorbate), and the propionates (sodium or calcium propionate), which are organic acids or their salts. Acidulants are used as preservatives because they increase the acidity of food, which can reduce growth of bacteria. Acidulants used include acetic acid, adipic acid, citric acid, fumaric acid, lactic acid, and phosphoric acid.

Processing Aids Substances used as manufacturing aids to enhance the appeal or utility of a food or food component, including clarifying agents, clouding agents, catalysts, flocculents, filter aids, and crystallization inhibitors.

Propane An aerating agent used in combination with chloropentafluoroethane or octafluorocyclobutane as a propellant and aerating agent for foamed or sprayed foods.

Propellants, Aerating Agents, and Gases Gases used to supply force to expel a product or used to reduce the amount of oxygen in contact with the food in packaging.

Propionic Acid The acid source of the propionates. Propionic acid in the liquid form has a strong odor and is corrosive, so it is used as the sodium, calcium, and potassium salts as a preservative. These yield the free acid in the pH range of the food in which they are used. It functions principally against mold. See **Calcium and Sodium Propionates.**

Propylene Glycol A humectant and flavor solvent that is a polyhydric alcohol (polyol). It is a clear, viscous liquid with complete solubility in water at 20°C and good oil solvency. It functions as a humectant, as do glycerol and sorbitol, in maintaining the desired moisture content and texture in foods such as shredded coconut and icings. It functions as a solvent for flavors and colors that are insoluble in water. It is also used in beverages and candy.

Propylene Glycol Alginate A gum that is the propylene glycol ester of alginic acid, which is obtained from kelp. As compared to sodium alginates, it has reduced sensitivity to acid and calcium salts. It functions in acidic systems. It functions as a thickener, stabilizer, and emulsifier in beer, salad dressings, syrups, and fruit drinks.

Propylene Glycol Ester See **Propylene Glycol Mono- and Di-Esters.**

Propylene Glycol Mono- and Di-Esters A liliphilic emulsifier that consists of propylene glycol esters of fatty acids, such as palmitic and stearic. It is used to increase the whipping ability and aeration in cake batters and whipped toppings.

Propylene Glycol Monostearate A lipophilic emulsifier that is a propylene glycol ester. It is used as a dispersing aid in nondairy creamers; as a crystal stabilizer in cake shortenings and whipped toppings; and as an aeration increaser in cake batters, icings, and toppings. It is also used in oils and shortenings.

Propyl 2-Furanacrylate A synthetic flavoring agent that is a stable, colorless to light yellow liquid of fruity odor. It should be stored in glass or tin-lined containers. It is used in flavors for apple, pear, and raspberry with applications in beverages, candy, and baked goods at 1 to 3 parts per million.

Propyl Gallate An antioxidant that is the *n*-propylester of 3,4,5-trihydroxybenzoic acid. Natural occurrence of propyl gallate has not been reported. It is commercially prepared by esterification of gallic acid with propyl alcohol followed by distillation to remove excess alcohol.

Propyl Hepatanoate A synthetic flavoring agent that is a stable, colorless liquid of fruity odor. It should be stored in glass, tin, or resin-lined containers. It is used in apple flavors and modified coffee. It has applications in beverages, ice cream, candy, and baked goods at 4 to 18 parts per million.

Propyl p-Hydroxybenzoate See **Parabens.**

Propyl Paraben See **Parabens.**

Psyllium A gum obtained from the plant of the *Plantago* genus. It hydrates slowly to form a viscous dispersion of concentrations up to 1 percent. A clear, gelatinous mass is formed at 2 percent. It is used in bulk laxatives.

Pyridoxine Vitamin B_6, a water-soluble vitamin with a solubility of 1 g in 5 ml of water. It functions in the utilization of protein and is an essential nutrient in enzyme reactions. It is necessary for proper growth. During processing, there is a loss due to leaching of the vitamin in water. It is destroyed by high temperatures, high irradiation, and exposure to light. During storage, loss increases with temperature and storage time. It is found in liver, eggs, and meats. It is also termed pyridoxine hydrochloride.

Pyridoxine Hydrochloride An acid form of Vitamin B_6, a water-soluble vitamin. It is soluble in water, and slightly soluble in alcohol. It is slowly affected by sunlight and is reasonably stable in air. It has a pH of 2.3 to 3.5. It is also termed vitamin B_6 hydrochloride. See **Pyridoxine.**

Q

Quicklime See **Calcium Oxide.**

Quince Seed A gum produced from the fruit of the quince tree *Cydonia oblonga.* It hydrates slowly to form a highly viscous dispersion at concentrations up to 1.5 percent. Above 2 percent, a slimy, muscilaginous mass is formed. It is principally used in the cosmetic industry. It is also termed gum quince seed, semen cydonia, golden apple seed, and cydonia seed.

Quinine A flavorant naturally obtained from the cinchona tree. It is used as a bitter flavoring in beverages such as quinine water, tonic water, and bitter lemon. Quinine sulfate and quinine hydrochloride are cleared for use as a flavor in carbonated beverages at levels less than 83 parts per million.

R

Raisin A dried grape used as a fruit and as an ingredient in cereals, baked goods, and desserts.

Raisin Seed Oil See **Grape Seed Oil.**

Rapeseed Oil The oil derived from seeds of *Brassica campestris* or *B. napus* of the family *Cruciferae* and related trees. It can function as a stabilizer and thickener in peanut butter and as an emulsifier in cake mix shortenings.

Rapeseed Oil, Fully Hydrogenated A stabilizer and thickener. A mixture of triglycerides in which the fatty acid composition is a mixture of saturated fatty acids. The fatty acids are present in the same proportions which result from the full hydrogenation of fatty acids occurring in natural rapeseed oil. Obtained from the *napus* and *compestris* varieties of *Brassica* of the family Cruciferae. Prepared by full hydrogenation of refined and bleached rapeseed oil at 310°F, using a catalyst such as nickel, until the iodine number is 4 or less. Used as a stabilizer and thickener in peanut butter.

Rapeseed Oil, Low Erucic Acid (Canola Oil) Fully refined, bleached, and deodorized oil obtained from certain varieties of *Brassica Napus* or *B. Campestris* of the family *Cruciferae.* Chemically, low erucic acid rapeseed oil is a mixture of triglycerides, composed of both saturated and unsaturated fatty acids, with an erucic acid content of no more than 2 percent of the component fatty acids. It may be partially hydrogenated to reduce the proportion of unsaturated fatty acids. Low erucic acid rapeseed oil and partially hydrogenated low erucic acid rapeseed oil are used in food, except in infant formula.

Rapeseed Oil, Superglycerinated Fully Hydrogenated An emulsifier that is a mixture of mono- and diglycerides with triglycerides as a minor component. The fatty acid composition is a mixture of saturated fatty acids present in the same proportions as those resulting from the full hydrogenation of fatty acids in a natural rapeseed oil. It is made by adding excess glycerol to the fully hydrogenated rapeseed oil and heating, in the presence of a sodium hydroxide catalyst, to 330°F under partial vacuum and steam sparging agitation. The ingredient is used as an emulsifier in shortenings for cake mixes.

Raw Sugar A sweetener that is an intermediate product, containing non-sugar impurities, thus being less refined than white sugar. It is made by crushing and shredding sugar cane to extract the juice which is processed to yield raw sugar and upon further processing yields refined cane sugar.

Red Durum Wheat Wheat obtained from the durum wheat kernel. It is used in macaroni and spaghetti products. See **Durum Wheat.**

Reduced Lactose Whey The portion of milk obtained by the removal of lactose from whey; the lactose content of the finished dry product does not exceed 60 percent. As with whey, reduced lactose whey can be used in fluid, concentrate, or a dry product form. The acidity of reduced lactose whey may be adjusted by the addition of safe and suitable pH-adjusting ingredients.

Reduced Minerals Whey The substance obtained by the removal of a portion of the minerals from whey; the dry product does not contain more than 7 percent ash. As with whey, reduced minerals whey can be used in fluid, concentrate, or a dry product form. The acidity of reduced minerals whey may be adjusted by the addition of safe and suitable pH-adjusting ingredients.

Reducing Sugar A sugar that can chemically react with copper in an alkaline solution. It combines with nitrogen compounds at elevated temperature to produce a browning "Maillard" reaction which contributes to the production of a brown crust in baked goods. It is used in the production of caramel color. Dextrose and fructose are reducing sugars.

Rennet A milk coagulant that is the concentrated extract of rennin enzyme obtained from calves' stomachs (calf rennet) or adult bovine stomachs (bovine rennet). The commercial saline extract of rennin contains a little pepsin, some sodium chloride, and some boric acid, sodium benzoate, or propylene glycol as a preservative. In the paste form, it also contains lipase. In the paste form it is used in Italian-type cheeses. It is used to coagulate milk into curd in making cheese and junket. A microbial rennet and a pepsin rennet also exist. See **Rennin.**

Rennet Casein The product that results from the precipitation of pasteurized milk with a rennet enzyme. Rennet casein requires a pH above 9 to dissolve, as compared to acid casein, which can be dissolved in alkali at a pH as low as 6.5. Rennet casein can be dispersed at lower pH by adding a complex phosphate such as sodium tripolyphosphate. This results in a casein of good emulsifying, whipping, foam stability, and water-binding properties. Uses include imitation cheese.

Rennin A milk coagulant that is an enzyme obtained from the abomasum portion of the stomach of suckling mammals. It is most active at pH 3.8. One part purified rennin will coagulate more than five million parts of milk. The commercial extract of rennin is termed rennet. It is used to coagulate milk in making cheese, junket, and custard. See **Rennet.**

Retinol The fat-soluble vitamin A which is required for new cell growth and prevention of night blindness. There is no appreciable loss by heating or freezing and it is stable in the absence of air. Sources include liver, fortified margarine, egg, and milk. Vitamin A palmitate can be found in frozen egg substitute.

Rhodinol A flavoring agent that is a colorless liquid, with an odor resembling rose. It is soluble in most fixed oils, mineral oils, mineral oil, and propylene glycol, insoluble in glycerin. It is usually obtained from reunion germanium oil.

Riboflavin The water-soluble vitamin B_2, required for healthy skin and the building and maintaining of body tissues. It is a yellow to orange-yellow crystalline powder. It acts as a coenzyme and carrier of hydrogen. It is stable to heat but may dissolve and be lost in cooking water. It is relatively stable to storage. Sources include leafy vegetables, cheese, eggs, and milk.

Rice Bran Oil An oil made from rice bran that consists mainly of oleic, linoleic, and palmitic fatty acids. It is used in salad oil, cooking oil, and hydrogenated shortenings.

Rice Bran Wax A refined wax obtained from rice bran. It is insoluble in water. It is used in candy, fresh fruits, and vegetables as a coating and as a plasticizing material in chewing gum.

Rice Flour The flour made from different varieties of long,- medium-, and short-grain rice, usually obtained from the broken milled rice. The chemical composition is the same as that of the whole rice. The flour does not contain gluten and, as a result, doughs made from it do not retain the gases generated during baking. Rice flours from different varieties display characteristic viscosity patterns during the heating and cooling of their pastes. In general, rice with starch of an amylose content greater than 22 percent has a relatively low peak viscosity and forms a rigid gel on cooling (high set-back viscosity). Rice with a starch low in amylose has a high peak and low set-back viscosity. Rice flour is used in formulated baby foods, breakfast foods, meat products, and breading.

Rice Starch The starch obtained from rice. It forms tender, opaque gels. It has some use in puddings.

Rochelle Salt See **Sodium Potassium Tartrate.**

Rosemary A spice made from the dried leaves of *Rosmarinus officinalis* L., an evergreen shrub. It has a medicinal, menthol flavor. It is available in whole and ground forms. It is used in soups, poultry, and meats, especially lamb.

Rum Ether A synthetic flavoring agent that is a stable, colorless to yellow liquid of ethereal rum-like note. It should be stored in glass and stainless steel containers. It is used to intensify rum flavors for application in beverages, candy, and ice cream at 67 to 320 parts per million and in alcoholic beverages at 1600 parts per million. It is also termed ethyl oxyhydrate.

Rye A cereal crop that is a source of rye flour. It is used as a bread grain.

Rye Flour The flour obtained by milling rye. It is available in white, medium, and dark grades and has a distinct flavor. It is usually diluted with wheat flour in order to make it more palatable. It is used in bread making.

S

Saccharin A non-nutritive synthetic sweetener which is 300 to 400 times sweeter than sucrose. It is nonhygroscopic and has a bitter aftertaste and a stability problem in cooked, canned, or baked goods. It is slightly soluble in water with a solubility of 10 g in 100 g of water at 25°C, but the solubility improves in boiling water. As sodium saccharin, there are two forms: 1,2-benzisothiazolin-3-one-1, 1-dioxide, sodium salt dihydrate, with a solubility of 1 g in 1.2 ml H_2O; and 1,2-benzisothiazolin-3-one-1,1-dioxide, sodium salt. Calcium saccharin (chemical name: 1,2-benzisothiazolin-3-one-1,1-dioxide, calcium salt) is used where low sodium content and reduced aftertaste are required. It is used in low-calorie foods such as jam, beverages, and desserts. It is also termed sodium benzosulfimide.

Sacharose See **Sugar.**

Safflower Oil An unsaturated oil obtained from the safflower seed of the plant *Carthamus tinctorius.* It consists mainly of linoleic and oleic fatty acids. It is used principally as a drying oil in the United States.

Saffron A spice obtained from the dried stigmas of the fall-flowering *Crocus-sativus* L. The flower stigma is of intense yellow color. It has a powerful, somewhat bitter aroma. It is used in breads, fish, chicken, sauces, and rice dishes.

Sage A spice made from the dried leaves of the shrub *Salvia officinalis* L. It has a strong, fragrant odor. It is available industrially as whole leaf, cut, rubbed, and ground to determined granulations. It is used in pork, soups, poultry seasonings, and fish.

Sago Starch The starch obtained from the sago palm *Metroxylon sagus* or *M. rumphii* and the palm fern *Cycas circinalis.* It forms high-strength gels which lose their clarity upon standing. It is used in confections and puddings.

Saint John's Bread See **Locust Bean Gum.**

Salt A seasoning and preservative whose chemical composition is sodium chloride, about 40 percent sodium and 60 percent chlorine by weight. It contains not less than 97.5 percent sodium chlroide after drying, while high-grade salt contains 99.8 percent sodium chloride. Salt production can be by solar evaporation, rock salt mining, and vacuum pan evaporation. The method selected depends on climate, character of the deposit, and type of salt required. Seasoned salt contains added flavors. It is available in several particle sizes (coarse, flake, fine) and shapes (flake, cube) which relate to density, solubility, flow, blending, and adherance. It is used as a carrier for dry or semidry ingredients or as an ingredient in prepared mixes. It is used in cheese, butter, and salted nuts for flavor. It is used in cheese manufacture to help remove the whey and suppress the growth of unwanted organisms, in sausage as a seasoning and curing agent, and in baked goods, pickles, and sauerkraut for flavor and fermentation control.

Santalol A flavoring agent that is a colorless or pale yellow quic, odor resembling sandalwood. It is soluble in alcohol, fixed oils, mineral oil, and propylene glycol; and insoluble in water and glycerin. It is obtained from a sandalwood oil source.

Savory A spice that is the dried leaves and flowering tops of the plant *Satureia hortenis* L. The two distinct varieties are summer savory and winter savory. Summer savory is generally preferred because it has a more delicate flavor and is less resinous. It is used in soups, salads, and sauces.

Sea Salt See **Salt.**

Seasoned Salt See **Salt.**

Self-Rising Flour White flour to which sodium bicarbonate and one or more of the acid-reacting substances are added, that is, monocalcium phosphate, sodium acid pyrophosphate, or sodium aluminum phosphate. It is seasoned with salt. The inclusion of these ingredients provides a leavening system that allows the flour to rise when wetted in the preparation of baked goods.

Semen Cydonia See **Quince Seeds.**

Semolina The purified ground middlings of durum wheat. It contains bran specks. Durum semolina is ground so that not more than 3 percent passes through a number 100 U.S. sieve. It takes longer to cook and is more resistant to overcooking than flour and results in less cloudiness in the water. It has a 50 percent relative protein efficiency as compared to nonfat dry milk. It is used in macaroni and spaghetti products.

Sequestrants Substances which combine with polyvalent metal ions to form a soluble metal complex, to improve the quality and stability of products. Examples include calcium citrate, calcium diacetate, calcium hexametaphosphate, citric acid, dipotassium phosphate, disodium phosphate, isopropyl citrate, monobasic calcium phosphate, monoisopropyl citrate, potassium citrate, sodium acid phosphate, sodium citrate, sodium gluconate, sodium hexametaphosphate, sodium metaphosphate, sodium phosphate, sodium pyrophosphate, sodium tripolyphosphate, stearyl citrate, and tetra sodium pyrophosphate.

Sesame Oil The oil obtained from sesame seeds. It consists principally of oleic and linoleic fatty acids. It has resistance to oxidation. It is used in vegetable shortenings, salad oil, and cooking oil, and is found in frozen chicken chow mein.

Sesame Seed The seed of the plant *Sesamum indicum* L. It has a sweet, "nutty" flavor. It yields sesame oil. It is used in breads, meats, and vegetables. It is also termed beene.

Shallot *Allium ascalonicum,* a member of the onion family. It ranges in size from walnut to small fig and is milder than the onion. It can be substituted for the onion and is used in sauces, dressings, soups, and meats.

Shortening Any animal or vegetable fat or oil that "shortens" or retards the development of gluten strands in baked goods for the purpose of producing a tender, crisp texture. Solid fats are most commonly used instead of oils because of their plastic nature. It is used in baked goods.

Silica, Amorphous See **Silicon Dioxide.**

Silicon Dioxide An anticaking agent, carrier, and dispersant that can absorb approximately 120 percent of its weight and remain free flowing. It is used in salt, flours, and powdered soups to prevent caking caused by moisture. It is also used in powdered coffee whitener, vanilla powder, baking powder, dried egg yolk, and tortilla chips. The usage level ranges from 1 to 2 percent. It is also termed silica, amorphous.

Skeletal Meat The edible part of the animal that is muscle tissue attached to the bone. It includes the shoulder and side of pork, brisket, flank, and round of beef. It is an ingredient in sausage.

Skim Milk Milk from which sufficient fat has been removed to reduce the milkfat content to less than 0.5 percent. It is used in the manufacture of certain cheese varieties, casein, and lactose. It is an ingredient in frozen deserts, baked goods, and confectionary. It is also consumed as a beverage.

Skim Milk Powder See **Milk Solids—Not-Fat.**

Slaked Lime See **Calcium Hydroxide.**

Smoke Flavoring A flavorant that can be obtained in the form of liquid smoke derived from burning hardwoods such as maple and hickory or as synthetic smoke made by synthesis. It is used to impart flavor and aroma to bacon, ham, and sausage.

Soda Alum See **Aluminum Sodium Sulfate.**

Sodium A metal element that.performs bodily functions.

Sodium Acetate A source of acetic acid that is obtained as crystals or powder. It has a solubility of 1 g in 0.8 ml of water.

Sodium Acetate, Anhydrous A source of acetic acid obtained as a granular powder. It has a solubility of 1 g in 2 ml of water.

Sodium Acid Carbonate See **Sodium Bicarbonate.**

Sodium Acid Phosphate See **Monosodium Phosphate.**

Sodium Acid Pyrophosphate A leavening agent, preservative, sequestrant, and buffer which is mildly acidic with a pH of 4.1. It is moderately soluble in water, with a solubility of 15 g in 100 ml at 25°C. It is used in doughnuts and biscuits for its variable gas release rate during the mixing, bench action, and baking process. It is used in baking powder as a leavening agent. It is used in canned fish products to reduce the level of undesired struvite crystals (magnesium ammonium phosphate hexahydrate) by complexing the magnesium. It is used to sequester metals in processed potatoes. It is also termed SAPP, disodium dihydrogen pyrophosphate, acid sodium pyrophosphate, disodium diphosphate, and sodium pyrophosphate.

Sodium Alginate A gum obtained as a sodium salt of alginic acid, which is obtained from seaweed. It is cold and hot-water soluble, producing a range of viscosities. It forms irreversible gels with calcium salts or acids. It functions as a thickener, binder, and gelling agent in dessert gels, puddings, sauces, toppings, and edible films.

Sodium Aluminosilicate See **Sodium Silicoaluminate.**

Sodium Aluminum Phosphate, Acidic A leavening agent, slowly soluble in water, which gives it a delayed leavening reaction. It has a pH of 2.8. Approximately 20 percent of the carbon dioxide is released during the mixing period and the remainder is released during the baking period when the batter is exposed to heat. It has a high tolerance to variation in batter preparation. It is used in prepared mixes such as cake and pancake mixes.

Sodium Aluminum Phosphate, Basic An emulsifier that is a white powder which is barely soluble in water. It has a pH of 9.2. It may be used in processed cheese to provide consistency and to aid in eliminating surface crystals.

Sodium Aluminum Sulfate A leavening agent that releases the majority of the gas during baking, and is not used alone but in combination with a faster-acting leavening agent such as monocalcium phosphate. This results in a double-acting baking powder. It is almost nonreactive until heat is applied. It is used in baked goods.

Sodium Ascorbate An antioxidant that is the sodium form of ascorbic acid. It is soluble in water and provides a non-acidic taste. A 10 percent aqueous solution has a pH of 7.3 to 7.6. In water, it readily reacts with atmospheric oxygen and other oxidizing agents, making it valuable as an antioxidant. One part sodium ascorbate is equivalent to 1.09 parts of sodium erythorbate. See **Ascorbic Acid.**

Sodium Benzoate A preservative that is the sodium salt of benzoic acid. It converts to benzoic acid, which is the active form. It has a solubility in water of 50 g in 100 ml at 25°C. Sodium benzoate is 180 times as soluble in water at 25°C as is the parent acid. The optimum functionality occurs between pH 2.5 to 4.0 and it is not recommended above pH 4.5. It is active against yeasts and bacteria. It is used in acidic foods such as fruit juices, jams, relishes, and beverages. Its use level ranges from 0.03 to 0.10 percent.

Sodium Benzosulfimide See **Saccharin.**

Sodium Bicarbonate A leavening agent with a pH of approximately 8.5 in a 1 percent solution at 25°C. It functions with food grade phosphates (acidic leavening compounds) to release carbon dioxide which expands during the baking process to provide the baked good with increased volume and tender eating qualities. It is also used in dry-mix beverages to obtain carbonation, which results when water is added to the mix containing the sodium bicarbonate and an acid. It is also termed baking soda and sodium acid carbonate.

Sodium Biphosphate See **Monosodium Phosphate.**

Sodium Bisulfite A preservative that exists as a powder, with a solubility of 1 g in 4 ml of water. It prevents discoloration and inhibits bacterial growth. It is used in dried fruit to inhibit browning and maintain the bright color. It is found in reconstituted lemon juice. See **Sulfur Dioxide.**

Sodium Calcium Aluminosilicate, Hydrated An anticaking agent for use at levels not to exceed 2 percent. It is also termed sodium calcium silicoaluminate.

Sodium Calcium Silicoaluminate See **Sodium Calcium Aluminosilicate, Hydrated.**

Sodium Caprate The sodium salt of capric acid. It functions as a binder, emulsifier, and anticaking agent.

Sodium Caprylate The sodium salt of caprylic acid. It functions as a binder, emulsifier, and anticaking agent.

Sodium Carbonate An alkali that exists as crystals or crystalline powder and is readily soluble in water. It has numerous functions: an antioxidant, a curing and pickling agent, a flavoring agent, a processing aid, a sequestrant, and an agent for pH control. It is used in instant soups to neutralize acidity. It is used in alginate water dessert gels to sequester the calcium, allowing the alginate to solubilize. It is also used in puddings, sauces, and baked goods.

Sodium Carboxymethylcellulose See **Carboxymethylcellulose.**

Sodium Carrageenan See **Carrageenan.**

Sodium Caseinate The sodium salt of casein, a milk protein. It is used as a protein source and for its functional properties such as water binding, emulsification, whitening, and whipping. It is used in coffee whiteners, nondairy whipped toppings, processed meat, and desserts.

Sodium Chloride See **Salt.**

Sodium Citrate A buffer and sequestrant obtained from citric acid as sodium citrate anhydrous and as sodium citrate dihydrate or sodium citrate hydrous. The crystalline products are prepared by direct crystallization from aqueous solutions. Sodium citrate anhydrous has a solubility in water of 57 g in 100 ml at 25°C, while sodium citrate dihydrate has a solubility of 65 g in 100 ml at 25°C. It is used as a buffer in carbonated beverages and to control pH in preserves. It improves the whipping properties in cream and prevents feathering of cream and nondairy coffee whiteners. It provides emulsification and solubilizes protein in processed cheese. It prevents precipitation of solids during storage in evaporated milk. In dry soups, it improves rehydration which reduces the cooking time. It functions as a sequestrant in puddings. It functions as a complexing agent for iron, calcium, magnesium, and aluminum. Typical usage levels range from 0.10 to 0.25 percent. It is also termed trisodium citrate.

Sodium Diacetate A preservative, sequestrant, acidulant, and flavoring agent that is a molecular compound of sodium acetate and acetic acid which yields acetic acid. It is a white crystalline powder which is hygroscopic. It functions against mold and bacteria and is used in bread.

Sodium Dioctylsulfosuccinate See **Dioctyl Sodium Sulfosuccinate.**

Sodium Erythorbate An antioxidant that is the sodium salt of erythorbic acid. In the dry crystal state it is nonreactive, but in water solution it readily reacts with atmospheric oxygen and other oxidizing agents, a property that makes it valuable as an antioxidant. During preparation, a minimal amount of air should be incorporated and it should be stored at a cool temperature. It has a solubility of 15 g in 100 ml of water at 25°C. On a comparative basis, 1.09 parts of sodium erythorbate are equivalent to 1 part of sodium ascorbate; 1.23 parts of sodium erythorbate are equivalent to 1 part erythorbic acid. It functions to control oxidative color and flavor deterioration in a variety of foods. In meat curing, it controls and accelerates the nitrite curing reaction and maintains the color brightness. It is used in frankfurters, bologna, and cured meats and is occasionally used in beverages, baked goods, and potato salad.

Sodium Ferric Pyrophosphate See **Sodium Iron Pyrophosphate.**

Sodium Ferrocyanide See **Yellow Prussiate of Soda.**

Sodium Hexametaphosphate A sequestrant and moisture binder that is very soluble in water but dissolves slowly. Solutions have a pH of 7.0. It permits peanuts to be salted in the shell by making it possible for the salt brine to penetrate the peanuts. In canned peas and lima beans, it functions as a tenderizer when added to the water used to soak or scald the vegetables prior to canning. It improves whipping properties in whipping proteins. It functions as a sequestrant for calcium and magnesium, having the best sequestering power of all the phosphates. It prevents gel formation in sterilized milk. It is also termed sodium metaphosphate, sodium polyphosphate, and Graham's salt.

Sodium Hydrogen Carbonate See **Sodium Bicarbonate.**

Sodium Hydrogen Diacetate See **Sodium Diacetate.**

Sodium Hydrogen Malate An acidulant.

Sodium Hydroxide An alkali that is soluble in water, having a solubility of 1 g in 1 ml of water. It is used to destroy the bitter chemicals in olives that are to become black olives. It also functions to neutralize acids in various food products.

Sodium Hypophosphite An emulsifier or stabilizer that is a white, odorless, deliquescent granular powder with a saline taste. It is also prepared as colorless, pearly crystalline plates. It is soluble in water, alcohol, and glycerol. It is prepared by neutralization of hypophosphorous acid or by direct aqueous alkaline hydrolysis of white phosphorus.

Sodium Hyposulfite See **Sodium Thiosulfate.**

Sodium Iron EDTA See **Iron.**

Sodium Iron Pyrophosphate A nutrient and dietary supplement that is a source of iron. It contains approximately 14.5 percent iron and is insoluble in water. It is utilized for the enrichment of foods that are susceptible to rancidity. It is also termed sodium ferric pyrophosphate.

Sodium Isoascorbate See **Sodium Erythorbate.**

Sodium Lactate A humectant that is the sodium salt of lactic acid which is low-melting and hygroscopic with a mildly saline taste. It is used in sponge cake and Swiss roll to produce a tender crumb and to reduce staling. It provides a protein plasticizing effect in biscuits. It is used in frankfurter-type sausages as a replacement for sodium chloride and as a dehydrating salt or humectant in uncured hams.

Sodium Laurate The sodium salt of lauric acid. It functions as a binder, emulsifier, and anticaking agent.

Sodium Lauryl Sulfate An emulsifier and whipping aid that has a solubility of 1 g in 10 ml of water. It functions as an emulsifier in egg whites. It is used as a whipping aid in marshmallows and angel food cake mix. It also functions to aid in dissolving fumaric acid.

Sodium Metabisulfite A preservative and antioxidant that exists as crystals or powder having a sulfur dioxide odor. It is readily soluble in water. It is used in dried fruits to preserve flavor, color, and to inhibit undesirable microorganism growth. It prevents "black spots" due to oxidative deterioration in shrimp. It is used in maraschino cherries. It is found in lemon drinks as a preservative. See **Sulfur Dioxide.**

Sodium Metaphosphate See **Sodium Hexametaphosphate.**

Sodium Myristate The sodium salt of myristic acid. It functions as a binder, emulsifier, and anticaking agent.

Sodium Nitrate The salt of nitric acid that functions as an antimicrobial agent and preservative. It is a naturally occurring substance in spinach, beets, broccoli, and other vegetables. It consists of colorless, odorless crystals or crystalline granules. It is moderately deliquescent in moist air and is readily soluble in water. It is used in meat curing to develop and stabilize the pink color. See **Nitrate.**

Sodium Nitrite The salt of nitrous acid that functions as an antimicrobial agent and preservative. It is a slightly yellow granular powder or nearly white, opaque mass or sticks. It is deliquescent in air. It has a solubility of 1 g in 1.5 ml of water. It is used in meat curing for color fixation and development of flavor. See **Nitrite.**

Sodium Oleate The sodium salt of oleic acid. It functions as a binder, emulsifier, and anticaking agent.

Sodium Palmitate The sodium salt of palmitic acid. It functions as a binder, emulsifier, and anticaking agent.

Sodium Pectinate See **Pectin.**

Sodium Phosphate, Dibasic See **Disodium Phosphate.**

Sodium Phosphate, Dibasic Dihydrate See **Disodium Phosphate.**

Sodium Phosphate, Monobasic See **Monosodium Phosphate.**

Sodium Phosphate, Tribasic See **Trisodium Phosphate.**

Sodium Polyphosphate A sequestrant and emulsifier that has a variety of functions in foods. It is used in cheese, dairy products, meat, fish, and poultry. It is specifically termed sodium tetrametaphosphate, sodium hexametaphosphate, and Graham's salt as determined by the chain length.

Sodium Potassium Tartrate · A buffer and sequestrant that is the salt of L(+)—tartaric acid. It has a solubility in water of 1 g in 1 ml. It is also termed Rochelle salt.

Sodium Propionate An antimicrobial agent that is the sodium salt of propionic acid. It occurs as colorless, transparent crystals or a granular crystalline powder. It is odorless or has a faint acetic-butyric acid odor, and is deliquescent. It is prepared by neutralizing propionic acid with sodium hydroxide. It is used in baked goods; nonalcoholic beverages; cheeses; confections and frostings; gelatins, puddings, and fillings; jams and jellies; meat products; and soft candy.

Sodium Pyrophosphate See **Sodium Acid Pyrophosphate.**

Sodium Pyrophosphate, Tetrabasic See **Tetrasodium Pyrophosphate.**

Sodium Saccharin See **Saccharin.**

Sodium Sesquicarbonate A pH control agent that is prepared by: (1) partial carbonation of soda ash solution followed by crystallization, centrifugation, and drying; (2) double refining of trona ore, a naturally occurring impure sodium sequicarbonate. It is used in cream manufacture at a level of the ingredient sufficient to control lactic acid prior to pasteurization and churning of cream into butter.

Sodium Silicate A product used as a preservative for eggs.

Sodium Silicoaluminate An anticaking and conditioning agent used to improve flow properties and prevent caking. It absorbs moisture up to 75 percent of its weight. It functions as a moisture absorbent, moisture barrier, carrier, and processing aid. It is used in salt, cake mixes, powdered sugar, nondairy creamers, and dry mixes. Usage level ranges from 1 to 2 percent. It is also termed sodium aluminosilicate.

Sodium Sorbate A preservative that is the salt of sorbic acid. It is partially soluble in water and is used effectively against yeasts and molds up to pH 6.5. It is not usually used as a replacement for sorbic acid or potassium sorbate. It is used in cheese and baked goods.

Sodium Stearate The sodium salt of stearic acid. It functions as a binder, emulsifier, and anticaking agent. It is used as a plasticizer in chewing gum base.

Sodium Stearoyl Fumarate A conditioning agent that functions as a dough conditioner for yeast-raised baked goods. It is also used as a conditioning agent for dehydrated potatoes.

Sodium Stearoyl Lactylate A dough conditioner, emulsifier, and whipping agent that is the reaction product of stearic and lactic acids neutralized to a sodium or calcium salt, for example, calcium stearoyl lactylate and sodium stearoyl lactylate. It is used to improve the tolerance of bread dough to processing and to improve gas retention. It is used as an emulsifier in coffee whiteners, puddings, and lowfat margarine. It functions as a whipping aid in egg products and vegetable fat toppings. It complexes starch in dehydrated potatoes to allow for production of thicker, more uniform sheets.

Sodium Stearyl Fumarate A dough conditioner and conditioning agent that is a white powder practically insoluble in water. It is used as a dough conditioner in yeast-raised baked goods. It is used as a conditioning agent in dehydrated potatoes. It also functions as a maturing and bleaching agent.

Sodium Sulfate The salt of sulfuric acid that is readily soluble in water and exists as crystals or crystalline powder. It is used in caramel production.

Sodium Sulfite See **Sulfur Dioxide.**

Sodium Tartrate A sequestrant and stabilizer that is the disodium salt of L(+)—tartaric acid. It is soluble in water. It functions as a sequestrant and stabilizer in meat products and sausage casings. It is also termed disodium tartrate.

Sodium Tetrametaphosphate A sequestrant and emulsifier that is infinitely soluble in water. It is used as a water binder in cured pork. It is also termed Graham's salt and sodium polyphosphate.

Sodium Thiosulfate A sequestrant, antioxidant, and formulation aid that is a powder soluble in water. It can be used in alcoholic beverages at 5 parts per million and in table salt at 0.1 percent. It is also termed sodium hyposulfite.

Sodium Triphosphate See **Sodium Tripolyphosphate.**

Sodium Tripolyphosphate A binder, stabilizer, and sequestrant that is mildly alkaline, with a pH of 10, and moderately soluble in water, with a solubility of 15 g in 100 ml of water at 25°C. It is used to improve the whipping properties of egg-containing angel food cake mix and meringues. It reduces gelling of juices and canned ham and tenderizes canned peas and lima beans. It is a moisture binder in cured pork and protects against discoloration and reduces shrinkage in sausage products. In algin desserts, it functions as a calcium sequestrant. It is also termed pentasodium tripolyphosphate and sodium triphosphate.

Softener A term used for ingredients that soften. Softening relates to the hygroscopicity and the ability of the polyhydric alcohol, such as propylene glycol or glycerin, to retain moisture. Softeners are used in shredded coconut, pet foods, and chewing gum to maintain moistness. It is also termed plasticizer.

Sorbic Acid A preservative that is effective against yeasts and molds. It is effective over a broad pH range up to pH 6.5, being ineffective above pH

7.0. It is a white, free-flowing powder which is slightly soluble in water with a solubility of 0.16 g in 100 ml of water at 20°C. Its solubility in water increases with increasing temperatures, although it is not recommended in foods that are pasteurized because it breaks down at high temperatures. The salts are potassium, calcium, and sodium sorbate. It is used in cheese, jelly, beverages, syrup, and pickles. Typical usage levels range from 0.05 to 0.10 percent.

Sorbitan Ester A lipophilic emulsifier whose permitted type in foods is sorbitan monostearate. It is used in cakes, chocolate, and coffee whitener. It is also termed sorbitan fatty acid ester.

Sorbitan Fatty Acid Ester See **Sorbitan Ester.**

Sorbitan Monostearate A lipophilic emulsifier that is a sorbitan fatty acid ester, being a sorbitol-derived analog of glycerol monostearate. It is a nonionic, oil-dispersible surface-active agent. It is used as a gloss enhancer in chocolate coatings; as a dispersant aid in coffee whiteners; to increase volume in cakes and icings; and often in combination with polysorbates. Typical usage level ranges from 0.30 to 0.70 percent.

Sorbitol A humectant that is a polyol (polyhydric alcohol) with good solubility in water and poor solubility in oil. It is approximately 60 percent as sweet as sugar. It maintains moistness in shredded coconut, pet foods, and candy. It is used in low-calorie beverages to provide body and taste. It is used in dietary foods such as sugarless candy and chewing gum. It is also used as a crystallization modifier in soft sugar-based confections.

Sorghum Oil An oil consisting mainly of linoleic and oleic fatty acids. It is similar in composition and properties to corn oil.

Soy Flour The powdered product obtained from defatted soybean. It has 50 percent or more protein content. It is used in doughnuts, bread, cereals, and sausage products as a nutrient and binder.

Soy Flour, Lecithinated See **Lecithinated Soy Flour.**

Soy Flour, Textured See **Textured Soy Flour.**

Soy Oil See **Soybean Oil.**

Soy Protein See **Soybean Protein.**

Soy Protein Concentrate See **Soybean Protein Concentrate.**

Soy Protein Isolate See **Soybean Protein Isolate.**

Soybean A legume of high protein content, containing 40 percent or greater protein and approximately 18 percent oil. The protein contains all the essential amino acids. Soybeans are processed to produce soybean flour, protein concentrate, protein isolate, and soybean oil.

Soybean Flour The flour made from defatted soybean, having a protein content in excess of 50 percent. It is used in doughnuts, cereal, bread, and sausage products for protein fortification and binding.

Soybean Oil The oil obtained from the seed of the soybean legume. It consists of approximately 86 percent unsaturated fatty acids with linoleic and oleic being the principal two fatty acids. It exists in hydrogenated and unhydrogenated forms. It is used in shortenings and margarine in the hydrogenated form. It has some use in salad and cooking oils in the unhydrogenated form, but is limited by its tendency to develop undesirable odor and flavor when in contact with air or when heated to frying temperatures.

Soybean Protein The protein obtained from soybeans, containing the essential amino acids. The most common forms are soybean flour (approximately 50 percent protein), soybean concentrate (approximately 70 percent protein), and soybean protein isolate (approximately 90 percent protein). It is used in sausages, snack foods, and meat analogs to provide emulsification, binding, moisture control, texture control, and protein fortification.

Soybean Protein Concentrate The concentrate obtained by processing soybean flour to remove the soluble carbohydrates. The protein content is approximately 70 percent. In the powder form, it is used in processed meat products and sausage products for moisture and fat binding as well as texture. In baby food, cereal, and snack food it provides protein fortification. In the granular form, it is used in ground meat food items for texture.

Soybean Protein Isolate The isolate prepared from soybean flour by extracting the protein and precipitating it to yield a product of approximately 90 percent protein. It functions to increase the protein content in foods, to reduce shrinkage, and to provide structure and appearance by emulsifying, stabilizing, and binding the fat and water. It is used in frozen spaghetti and meatballs, whipped toppings, and snack foods.

Spice A variety of dried plant products that exhibit an aroma and flavor and from which no volatiles or other flavoring principles have been removed.

Spirit Vinegar The product made by the acetous fermentation of dilute distilled alcohol, containing not less than 4 g of acetic acid per 100 cm^3 at 20°C. It functions as an acidulant and provider of flavor. It is used in mayonnaise, sauces, and salad dressings. It is also termed distilled vinegar and grain vinegar.

Stabilizers and Thickeners Substances used to produce viscous solutions or dispersions, to impart body, improve consistency, or stabilize emulsions, including suspending and bodying agents, setting agents, jellying agents, and bulking agents, etc.

Stannous Chloride An antioxidant and preservative that exists as white or colorless crystals, being very soluble in water. It reacts readily with oxygen, preventing its combination with chemicals and foods which would otherwise result in discoloration and undesirable odors. It is used for color retention in asparagus at less than 20 parts per million. It is also used in carbonated drinks.

Starch A carbohydrate consisting of glucose units containing amylose and amylopectin which contribute to varying starch properties. Starch is insoluble in cold water, but upon heating the starch granules swell and burst forming starch paste. Starch sources include arrowroot, corn, potato, rice, sago, tapioca, waxy corn, and wheat. Starches are modified by treatment to alter their functional properties. Terminology designating these starches includes acid-modified corn starch, food starch modified, modified food starch, oxidized corn starch, pregelatinized starch, thin-boiling starch, and wheat starch, gelatinized. See specific starch.

Stearic Acid A fatty acid that is a mixture of solid organic acids obtained principally from stearic acid and palmitic acid. It is practically insoluble in water. It functions as a lubricant, binder, and defoamer. It is used as a softener in chewing gum base.

Stearoyl Lactylate A dough conditioner, emulsifier, and whipping agent that is the reaction product of stearic and lactic acid neutralized to a sodium or calcium salt, for example, calcium stearoyl lactylate and sodium stearoyl lactylate. It is used to improve the tolerance of bread dough to processing and to improve gas retention. It is used as an emulsifier in coffee whiteners, puddings, and lowfat margarine. It functions as a whipping aid in egg products and vegetable fat toppings. It complexes starch in dehydrated potatoes to allow for production of thicker, more uniform sheets.

Stearoyl Propylene Glycol Hydrogen Succinate See **Succistearin.**

Stearyl Citrate An antioxidant made by reacting citric acid, which is not soluble in fats and oils, with stearyl alcohol, which readily dissolves in oils, thus enabling the citrate to dissolve in oil. It prevents metal ions from catalyzing oxidative reactions which cause rancidity. It is related to isopropyl citrate. It is used in vegetable oils and margarines.

Stearyl Monoglyceridyl Citrate An emulsion stabilizer prepared by controlled chemical reaction of citric acid, monoglycerides of fatty acids, and stearyl alcohol. It is used in or with shortenings containing emulsifiers.

Sterculia Gum See **Karaya.**

Straight Flour All of the flour that can be milled from a wheat blend, or approximately 72 percent of the wheat kernel which equates to 100 percent separation.

Succinic Acid An acidulant that is commercially prepared by the hydrogenation of maleic or fumaric acid. It is a nonhygroscopic acid but is more soluble in 25°C water than fumaric and adipic acid. It has low acid strength and slow taste build-up; it is not a substitute for normal acidulants. It combines with proteins in modifying the plasticity of bread dough. It functions as an acidulant and flavor enhancer in relishes, beverages, and hot sausages.

Succinic Anhydride An acidulant that hydrolyzes very slowly to succinic acid in water. It has thermal stability and a low melting point (118°C) which permits it to be used in products at comparatively low temperatures. It is used as a leavening acidulant for baking powder.

Succinylated Monoglycerides Emulsifiers and dough conditioners made by the dissociation of succinylated monoglycerides. They are used in baked goods at 0.056 to 0.113 kg per 45.4 kg of flour to provide dough strength, improve shelf life, and improve texture. They are also used in shortenings.

Succistearin (Stearoyl Propylene Glycol Hydrogen Succinate) An emulsifier that is the reaction product of succinic anhydride, fully hydrogenated vegetable oil (predominantly C16 or C18 fatty acid chain length), and propylene glycol. It is used in or with shortenings and edible oils intended for use in cakes, cake mixes, fillings, icings, pastries, and toppings.

Sucrose See **Sugar.**

Sucrose Fatty Acid Esters Emulsifiers, texturizers that are the mono-, di-, and tri-esters of sucrose with fatty acids and are derived from sucrose and edible tallow, or hydrogenated edible tallow or edible vegetable oils. Ethyl acetate or methyl ethyl ketone or dimethyl sulfoxide and isobutyl alcohol (2-methyl-1-propanol) may be used in the preparation of sucrose fatty acid esters. Sucrose fatty acid esters may be used as follows: as emulsifiers in baked goods and baking mixes, in dairy product analogs, in frozen dairy desserts and mixes, and in whipped milk products; as texturizers in biscuit mixes; as components of protective coatings applied to fresh apples, avocados, bananas, banana plantains, limes, melons (honeydew and cantaloupe), papaya, peaches, pears, pineapples, and plums to retard ripening and spoiling.

Sugar A sweetener that is the disaccharide sucrose, consisting of one molecule of glucose and one molecule of fructose. It is obtained as cane or beet sugar. It has relatively constant solubility and is a universal sweetener because of its intense sweetness and solubility. It is available in various forms which include granulated, brown, and powdered. It is used in desserts, beverages, cakes, ice cream, icings, cereals, and baked goods. It is also termed beet sugar, cane sugar, and sucrose.

Sugar Beet Extract Flavor Base A flavor that is the concentrated residue of soluble sugar beet extractives from which sugar and glutamic acid have been recovered, and which has been subjected to ion exchange to minimize the concentration of naturally occurring trace minerals. It is used as a flavor in food.

Sugar, Brown See **Brown Sugar.**

Sugar, Natural See **Turbinado Sugar.**

Sugar, Powdered See **Powdered Sugar.**

Sugar, Raw A natural sugar that has been washed to remove the impurities. It has a light golden color resulting from the molasses and a larger crystal size than granulated sugar. It is used where the flavor of natural sugar is desired, such as in cookies, bread, and cakes.

Sugar, Reducing See **Reducing Sugar.**

Sugar Syrup A sweetener that is clear solutions of sucrose existing in varying grades. There is a water-white grade which is a sparkling clear syrup used in canned goods and beverages. There is also a light straw grade which has small amounts of color and nonsugar components.

Sugar Syrup, Invert See **Invert Sugar Syrup.**

Sugar, Washed Raw See **Turbinado Sugar.**

Sulfur Dioxide A preservative, being a gas that dissolves in water to yield sulfurous acid. Sulfite salts, such as sodium and potassium sulfite, sodium and potassium bisulfite, and sodium and potassium metabisulfite, yield free sulfurous acid at low pH. Sulfur dioxide prevents the discoloration of foods by combining with the sugars and enzymes. It also inhibits bacterial growth. It is used in beverages, cherries, wines, and fruits.

Sulfuric Acid An acidulant that is a clear, colorless, odorless liquid with great affinity for water. It is prepared by reacting sulfur dioxide with oxygen and mixing the resulting sulfur trioxide with water, or by reacting nitric oxide with sulfur dioxide in water. It is very corrosive. It is used as a modifier of food starch and is used in caramel production and in alcoholic beverages.

Sunflower Oil A highly polyunsaturated oil obtained from sunflower seeds. There are two types of sunflower grown: an oilseed type used as a vegetable oil, and a non-oilseed type used for human food and bird seed. The composition of sunflower oil varies according to location and growing temperature. In general, sunflowers grown above the 39th parallel are high in linoleic acid and those grown below are high in oleic acid. The high linoleic variety is used for margarine and salad oil, while the high oleic variety is used in frying applications. This bland-flavored oil has a smoke point of 485 to 490°F (252–254°C) which gives it utility in baking, cooking, and frying foods. It is also used as a salad oil. In the hydrogenated form, it is used in margarine and shortenings.

Sunset Yellow FCF See **FD&C Yellow #6.**

Surface-Active Agents Agents used to modify surface properties of liquid food components for a variety of effects, other than emulsification. Agents include: solubilizing agents, dispersants, detergents, wetting agents, rehydration enhancers, whipping agents, foaming agents, and defoaming agents.

Surface-Finishing Agents Agents used to increase palatability, preserve gloss, and inhibit discoloration of foods, including glazes, polishes, waxes, and protective coatings. Examples include: coumarone-indene resin, methyl esters of fatty acids produced from e fats and oils, microcapsules for flavoring substances, morpholine, oxidized polyethylene, petroleum naphtha, polyacrylamide, sulfated butyl oleate, synthetic paraffin and succinic derivatives, terpene resin.

Suspending Agents See **Gums.**

Sweeteners These can be classified as natural or artificial. The natural sweeteners are carbohydrates consisting of molecules of carbon, hydrogen, and oxygen. The simplest form of carbohydrate is the monosaccharide or simple sugar and includes glucose (dextrose), fructose (levulose), and galactose, which are six-carbon (hexose) sugars. An example of an artificial sweetener is aspartame.

Sweet Pepper See **Paprika.**

Sweet Rice Flour See **Waxy Rice Flour.**

Synthetic Glycerin Produced by Hydrogenolysis of Carbohydrates An emulsifier produced by the hydrogenolysis of carbohydrates may be safely used in food. It contains equal to or less than 0.2 percent by weight of a mixture of butanetriols.

Synthetic Petroleum Wax A wax that is a mixture of solid hydrocarbons, parafinic in nature, prepared by catalytic polymerization of ethylene. Synthetic petroleum wax is used in chewing gum base as a masticatory substance, on cheese and raw fruits and vegetables as a protective coating, and as a defoamer in food.

T

Tallow Animal fat obtained by separation from connected tissue, usually in mutton or beef. It consists principally of oleic and palmitic acid. It is a source of fat and is used in cake mix. It is used mostly in shortening and cooking oils.

Tangerine Oil, Expressed A flavoring agent that is a red, brown, or orange oily liquid with a pleasant aroma. Oil obtained from unripe fruit may be green. It is soluble in most fixed oils and mineral oil, slightly soluble in propylene glycol, insoluble in glycerin. It is obtained by expression of oils from peels of ripe fruit of Dancy tangerine and related varieties.

Tannic Acid A sequestrant that refers to a mixture of hydrolyzable tannins of a more complex structure than gallic acid. It is used in clarifying beer and wine. See **Tannins.**

Tannins These are phenolic compounds that have several hydrolyzable groups. They are classified as: (a) hydrolyzable, yielding phenols such as gallic acid in the presence of acid and heat; and (b) condensed, obtained from the extract of oak trees and not hydrolyzable. Tannins are used for taste and chemical properties and as a sequestrant. They affect the color and flavor of fruits and vegetables. They are used in fruits, wine, and beer to remove undesirable material by forming insoluble complexes with the proteins.

Tapioca Starch Starch having a bland flavor, being opaque, and forming long, cohesive pastes. It is found mainly in the modified form, being the pearl, granulated form which has been treated to be less stringy. It is used in puddings and pie fillings.

Taro The tropical tuber *Colocasia esculenta* which can be used to make poi. Poi is carbohydrate food made by cooking the underground stem (corm) of the taro plant. The corms must be cooked because the calcium oxalate crystals present in the raw vegetable will act as tiny needles in the mouth.

Tarragon The dried leaves and flowering tops of the herb *Artemisia dracunculus* L. It has a distinct aroma and anise-like flavor. It is used in salads, fish, sauces, and vinegar.

Tartaric Acid An acidulant that occurs naturally in grapes. It is hygroscopic and rapidly soluble, with a solubility of 150 g in 100 ml of distilled water at 25°C. It has a slightly tarter taste than citric acid, with a tartness equivalent of 0.8 to 0.9. It augments the flavor of fruits in which it is a natural constituent. It is used in grape- and lime-flavored beverages and grape-flavored jellies. It is used as an acidulant in baking powder and as a synergist with antioxidants to prevent rancidity.

Tartrazine See **FD&C Yellow #5.**

Terpene Resin A moisture barrier that is the betapinene polymer obtained by polymerizing terpene hydrocarbons derived from wood. It is used on soft gelatin capsules in an amount not to exceed 0.07 percent of the weight of the capsule, and on powders of ascorbic acid or its salts in an amount not to exceed 7 percent of the weight of the powder.

Tertiary Butylhydroquinone (TBHQ) An antioxidant that exhibits an excellent stabilizing effect in unsaturated fats and oils. It has good solubility in fats and oils, with a maximum usage level of 0.02 percent based on the weight of the fat or oil or the fat content of the food product. It shows no discoloration in the presence of iron and produces no discernible flavor or odor. It can be combined with BHA and BHT. It is used in edible fats and vegetable oils to retard rancidity. It is used in potato chips and dry cereal. It is also termed mono-tertiary-butylhydroquinone.

Tetrahydrofurfuryl Acetate A synthetic flavoring agent that is a stable, colorless liquid of slightly fruity odor. It should be stored in glass or tin containers. It is used in fruit flavors for application in candy and baked goods at 2 to 20 parts per million.

Tetrahydrofurfuryl Propionate A synthetic flavoring agent that is a stable, colorless liquid of chocolate note. It should be stored in glass or tin-lined containers. It is used in flavors for chocolate with applications in beverages and ice cream at 2 parts per million, and in candy and baked goods at 20 parts per million.

Tetrasodium Diphosphate See **Tetrasodium Pyrophosphate.**

Tetrasodium Pyrophosphate A coagulant, emulsifier, and sequestrant that is mildly alkaline, with a pH of 10. It is moderately soluble in water, with a solubility of 0.8 g per 100 ml at 25°C. It is used as a coagulant in noncooked instant puddings to provide thickening. It functions in cheese to reduce the meltability and fat separation. It is used as a dispersant in malted milk and chocolate drink powders. It prevents crystal formation in tuna. It is also termed sodium pyrophosphate, tetrasodium diphosphate, and TSPP.

Textured Soy Flour Soy flour that is processed and extruded to form products of specific texture and form, such as meatlike nuggets. The formed products are crunchy in the dry form and upon hydration become moist and chewy.

Textured Vegetable Protein A vegetable protein that is processed and extruded to form beeflike strips, meatlike nuggets, or other analogs. In the dehydrated form, the analogs are crunchy and upon hydration become moist and chewy. Soy protein is the most popular protein source, but other vegetable proteins include peanut and wheat. It is used as meat analogs. It is also termed textured soy flour or textured soy protein.

Thiamine The water-soluble vitamin B_1, required for normal digestion and functioning of nerve tissues and in the prevention of beriberi. It also acts as a coenzyme in the metabolism of carbohydrates. During processing, the higher and longer the heating period, the greater the loss. The loss is reduced in the presence of acid. Thiamine hydrochloride and thiamine mononitrate are two available forms. The mononitrate form is less hygroscopic and more stable than the hydrochloride form, making it suitable for use in beverage powders. It is used in enriched flour and is found as thiamine mononitrite in frozen egg substitute and crackers.

Thiamine Mononitrate See **Thiamine.**

THBP An antioxidant (2,4,5-trihy-droxybutyrophenone) that is used alone or in combination with other permitted antioxidants. The total antioxidant content of a food containing the additive will not exceed 0.02 percent of the oil or fat content of the food, including the essential (volatile) oil content of the food.

Thin-Boiling Starch See **Cornstarch, Acid-Modified.**

Thiodipropionic Acid An antioxidant used to prevent fats and oils from going rancid. It has the same functionality as BHA, BHT, and propyl gallate.

Thyme The dried leaves and flowering tops of the shrub *Thymus vulgaris* L. There are two important variations: French thyme, which has a narrow leaf; and lemon thyme, which has a variegated leaf. It is used in soups, cheese, sauces, and appetizers.

Titanium Dioxide A white pigment that disperses in liquids and possesses great opacifying power. The crystalline modifications of titanium dioxide are rutile and anatase, of which only anatase finds use as a color additive.

Tocopherol Fat-soluble vitamin E, which is a light yellow oil readily degradable by heat. As a vitamin, it is essential for normal muscle growth and prevents vitamin A destruction by deterioration. It also functions as an antioxidant. It prevents the oxidation of certain fatty acids and is stable unless the food becomes rancid. Vegetable oils contain a higher concentration of natural antioxidants, including tocopherols, than animal fats and are thus more stable. Tocopherol is obtained from vegetable oils, beans, eggs, and milk.

Tofu A soybean curd product. Soybeans are soaked, ground, and filtered, with the remaining mixture being heated to 75°C and a coagulant added, which results in the formation of the soy curd and whey. The soy curd is pressed to separate it from the whey and then washed and cooled. It is low in calories and saturated fats while high in vitamins, minerals, and digestible protein. It is tasteless, but takes on the flavors of the products with which it is cooked. Uses include frozen desserts and meat products.

Tomato Paste The paste prepared from tomatoes which are processed by heat to prevent spoilage. The paste contains not less than 24 percent tomato soluble solids.

Tragacanth A gum produced from a bush of the genus *Astragalus.* It swells in water to give a highly viscous sol or paste. A 1 percent solution of the purest gum has a viscosity of approximately 3400 cps, and about 2 percent can form a paste. Solutions have a pH of 5 to 6. It is stable at low pH and is an effective suspending agent because of its stability and acid resistance. It is used in salad dressings, sauces, fruit fillings, and citrus beverages.

Triacetin See **Glyceryl Triacetate.**

Tribasic Calcium Phosphate See **Tricalcium Phosphate.**

Tributyrin (Butyrin or Glyceryl Tributyrate) A flavoring agent that is the triester of glycerin and butyric acid. It is prepared by esterification of glycerin with excess butyric acid. It is used in the following foods: baked goods; alcoholic beverages; nonalcoholic beverages; fats and oils; frozen dairy desserts and mixes; gelatins, puddings and fillings; and soft candy.

Tricalcium Orthophosphate See **Tricalcium Phosphate.**

Tricalcium Phosphate An anticaking agent and calcium source that is a white powder that is almost insoluble in water. It is used as an anticaking agent in table salt and dry vinegar. It is used as a source of calcium and phosphorus in cereals and desserts. It functions as a bleaching agent in flour and in lard, and prevents undesirable coloring and improves stability for frying. It is also termed tribasic calcium phosphate, tricalcium orthophosphate, and precipitated calcium phosphate.

Tricalcium Silicate An anticaking agent used in table salt.

Triethyl Citrate A sequestrant that is an oily liquid, slightly soluble in water. It is found in lemon drinks.

Trihydroxybutyrophenone (THBP) An antioxidant. See **THBP.**

Tripotassium Citrate A buffer and sequestrant that possesses the advantageous properties of citric acid but without the acid reactions. See **Potassium Citrate, Monohydrate.**

Trisodium Citrate A buffer and sequestrant that is the trisodium salt of citric acid. See **Sodium Citrate.**

Trisodium Monophosphate See **Trisodium Phosphate.**

Trisodium Orthophosphate See **Trisodium Phosphate.**

Trisodium Phosphate An emulsifier and buffer that is strongly alkaline, with a pH of 12. It is moderately soluble in water, with a solubility of 14 g per 100 ml at 25°C. It functions as an emulsifier in processed cheese to improve texture. It maintains viscosity and prevents phase separation in evaporated milk and is also found in cereals. It is also termed trisodium orthophosphate, sodium phosphate tribasic, and trisodium monophosphate.

Trisodium Phosphate Crystals An emulsifier and buffer with a solubility in water of 50 g per 100 ml at 25°C. It is used in processed cheese as an emulsifier and it is also used in denture cleaner formulations.

Turbinado Sugar Washed raw sugar of light gold color and larger grain size than regular sugar. It has a thin film of molasses which contributes toward the distinctive flavor. It is also termed natural sugar and washed raw sugar.

Turmeric A spice and colorant that is the rhizome or root of *Curcuma longa*. As a spice, it has a taste related to mustard. As a vegetable color, it has a bright yellow to greenish-yellow hue. The yellow pigment is curcumin. It is water miscible and has excellent heat stability, poor light and pH stability, and good tinctorial strength. It exists as an extract and oleoresin. It is used in meat, poultry, fish, and rice dishes.

U

(Gamma)-Undecalactone A synthetic flavoring agent that is a colorless to yellow liquid of strong peach fruit odor. It is unstable to alkali and stable to weak organic acids. It should be stored in glass or tin containers. It is used in flavors for its peach note and has application in gelatins, puddings, beverages, ice cream, and candy at 7 to 11 parts per million.

Undecanal (Aldehyde C-11 Undecyclic; n-Undecyl Aldehyde) A flavoring agent that is a liquid, colorless, or pale-yellow, with a sweet odor. It is soluble in most fixed oils, mineral oil, and propylene glycol; insoluble in glycerin. It is obtained by chemical synthesis.

V

Vanaspati A vegetable fat used in candy.

Vanilla A flavorant obtained from the cured vanilla bean. The vanilla or vanilla bean refers to the fully grown, unripe, cured and dried fruit pod of the vanilla vine *Vanilla planifolia*. Those beans produced in Madagascar and its neighboring islands are termed "Bourbon beans"; those produced in Indonesia are termed "Java beans." The bean contains 1.5 to 3.0 percent vanillin, the most powerful flavorant in the cured bean, along with approximately 10 percent of other extractives. It is used in the comminuted form in "Philadelphia" type ice cream or as a vanilla flavorant in sauces or liquids by suspending the whole bean in them. Most vanilla flavoring is done with vanilla extract.

Vanilla Extract A flavorant made from vanilla bean extract. It is a solution containing not less than 35 percent alcohol of the components extracted from one or more units of vanilla constituent. One unit is 0.378 kg of vanilla beans containing not more than 25 percent moisture. A double-strength solution (2-fold) contains twice the quantity of beans. It is used in desserts, baked goods, and beverages.

Vanilla Flavor, Artificial A flavorant that consists of vanilla reinforced with synthetic vanillin. The best imitation vanillas contain vanillin, ethyl vanillin, or very little coumarin with or without vanilla, while the poorer ones contain high levels of coumarin. It is used in desserts, baked goods, and beverages.

Vanilla Sugar A flavorant consisting of sugar mixed with vanilla extract. It is used in desserts and other food products.

Vanillin A flavorant made from synthetic or artificial vanilla which can be derived from lignin of whey sulfite liquors and is synthetically processed from guaiacol and eugenol. The related product, ethyl vanillin, has 3.5 times the flavoring power of vanillin. Vanillin also refers to the primary flavor ingredient in vanilla, which is obtained by extraction from the vanilla bean. Vanillin is used as a substitute for vanilla extract, with application in ice cream, desserts, baked goods, and beverages at 60 to 220 parts per million.

Vanillin Acetate A synthetic flavoring agent that is moderately stable, white to yellow crystals of vanilla odor. It should be stored in glass or polyethylene-lined containers. It is used in flavors for vanilla note, with application in beverages, ice cream, candy, and baked goods at 11 to 28 parts per million.

Vegetable Gums Gums that are water thickeners obtained from a plant source.

Vegetable Oils Oils obtained from a vegetable source, including soy beans, peanuts, cottonseeds, and palms. They are used in cooking and salad oils and dressings.

Vegetable Oil, Hydrogenated See **Hydrogenated Vegetable Oil.**

Vegetable Protein, Textured See **Textured Vegetable Protein.**

Vinegar An acidulant and flavorant that, with regard to general types, is the product produced from cider, grapes, sucrose, glucose, or malt by successive alcoholic and acetous fermentations in which acetic acid is the principal measured component. The term *vinegar* applies only to cider vinegar, also termed apple vinegar. The acetic acid content is measured in grains, where 10 grains equals 1 percent acetic acid. It is used in salad dressing, dressings, and sauces.

Vinegar, Distilled The product made by the acetous fermentation of dilute distilled alcohol without addition of color, containing not less than 4 g of acetic acid per 100 cm^3 at 20°C. The acetic acid content is measured in grains, where 10 grains equal 1 percent acetic acid. It is used in mayonnaise and salad dressing. It is also termed spirit vinegar and grain vinegar.

Vital Wheat Gluten A powder of high protein content obtained by drying freshly washed gluten under controlled-temperature conditions. It absorbs approximately twice its weight of water and readily forms a cohesive, elastic dough. It is used in bread, rolls, and buns. See **Wheat Gluten.**

Vitamins Organic compounds that are essential for normal body growth and maintenance. They are classified into two groups: fat-soluble (vitamins A, D. E, and K), and water-soluble (vitamins B and C). Vitamins are measured in very low concentrations, such as 1 to 100 mg. Through biochemical action, they perform various functions in such processes as cell growth, normal digestion, manufacture of red blood cells, and absorption of calcium and phosphorus. Inadequate vitamin intake can be the result of food deficiency, increased vitamin requirements, and increased vitamin loss. The vitamins of determined importance include: A (retinol), B_1 (thiamine), B_2 (riboflavin), B_5 (pantothenic acid), B_6 (pyridoxime), B_{12} (cyanocobalamin), C (ascorbic acid), D (calciferol), E (tocopherol) K, niacin, folic acid, and biotin.

Vitamin A See **Retinol.**

Vitamin B_1 See **Thiamine.**

Vitamin B_2 See **Riboflavin.**

Vitamin B_5 See **Pantothenic Acid.**

Vitamin B_6 See **Pyridoxine.**

Vitamin B_6 Hydrochloride See **Pyridoxine Hydrochloride.**

Vitamin B_{12} See **Cyanocobalamin.**

Vitamin C See **Ascorbic Acid.**

Vitamin D_2 See **Calciferol.**

Vitamin E See **Tocopherol.**

Vitamin K A fat-soluble vitamin that is essential for blood clotting. It is destroyed by irradiation during processing but has no appreciable loss during storage. It occurs in spinach, cabbage, liver, and wheat bran.

Washed Raw Sugar A raw sugar of larger grain size. See **Turbinado Sugar.**

Water A colorless, odorless, tasteless liquid formed by the combination of two hydrogen and one oxygen atoms. It allows substances to dissolve and functions as a solvent, dispersing medium, hydrate, and promoter of chemical changes. It is a major constituent in meats, fruits, and vegetables. Distilled water is obtained by condensation of water vapor.

Waxy Corn A corn consisting essentially of amylopectin (pure branched-chain polymers), which differentiates it from regular corn, which consists of amylose and amylopectin. The amylopectin content results in a starch which upon heating forms a clear, cohesive paste that does not form a true gel upon cooling. It has a high water-binding capacity and is resistant to gel formation and retrogradation. It is used in puddings and sauces.

Waxy Maize Starch The starch portion of waxy corn, consisting essentially of amylopectin. It yields pastes that are almost clear upon cooling and are noncongealing. It forms a translucent, water-soluble coating when dried in thin films. It is used to thicken a variety of foods such as sauces and puddings. It is also termed waxy starch and amioca.

Waxy Rice Flour A flour obtained from waxy rice, which contains almost no amylose. It is comparable in viscosity characteristics to waxy corn flour. It has less than 0.5 percent amylose in the starch and contains alpha-amylose. It has excellent resistance to syneresis during freeze/thaw cycles. It is used in frozen sauces and gravies. It is also called sweet rice flour.

Waxy Rice Starch A rice starch that contributes freeze-thaw stability to sauces and puddings but may provide objectionable flavor.

Waxy Sorghum A type of sorghum characterized by having paste clarity, high water-binding capacity, and resistance to gel formation and retrogradation. The unmodified form results in a stringy, cohesive paste. It is used in dressings with other starches.

Waxy Starch See **Waxy Maize Starch.**

Wetting Agents See **Surface-Active Agents.**

Wheat A cereal grain in which the kernel is separated by milling into flour, bran, and germ. It is used in all types of farinaceous foods. See **Flour category.**

Wheat Flour A fine powdery substance obtained by milling wheat with application in farinaceous foods.

Wheat Germ The oil-containing portion of the wheat kernel.

Wheat Gluten The water-insoluble complex protein fraction separated from wheat flours. Gum gluten is wheat gluten in its freshly extracted wet form. Dry gluten is approximately 70 to 80 percent protein but is deficient in the amino acid lysine. It absorbs two to three times its weight in water. The differences in properties of wheat gluten in comparison to almost all other food proteins are largely due to the low polarity level of the total amino acid structure. Most food proteins have polar group levels of 30 to 45 percent and have a net negative charge, while wheat gluten has a polar group level of approximately 10 percent with a net positive charge. This results in the repulsion of excess water and the close association of the wheat gluten molecules and resistance to dispersion. In baked goods, this results in the ability to form adhesive, cohesive masses, films, and three-dimensional networks. Gluten formation is utilized in the baking industry to impart dough strength, gas retention, structure, water absorption, and retention with breads, cakes, doughnuts, and so on. It is also used as a formulation aid, binder, filler, and tableting aid. See **Gluten.**

Wheat Starch A starch obtained from wheat. It produces lower viscosity and more tender gels than starch obtained from corn or sorghum. It has a gelatinization range about 10°C lower than corn or waxy maize starch. It is used in the baking industry to permit the use of hard wheat flour in baked goods. It functions as a binder in breading and batter mixes. It is used in soups, pie fillings, sauces, and gravies.

Wheat Starch, Gelatinized See **Pregelatinized Starch.**

Whey The portion of milk remaining after coagulation and removal of curd. There are two principal types: sweet whey obtained during the making of rennet-type hard cheeses like Cheddar and Swiss, with a pH of approximately 6.1; and acid whey obtained during the making of acid-type cheeses, such as cottage cheese, with a pH of approximately 4.4 to 4.6. Whey is used as a source of lactose, milk solids, and whey proteins. It is used in baked goods, ice cream, and dry mixes.

Whey Protein Concentrate The dry portion of whey obtained by the removal of sufficient nonprotein constituents from whey so that the finished dry product contains not less than 25 percent protein. As with whey, whey protein concentrate can be used in fluid, concentrate, or dry product form. The acidity of whey protein concentrate may be adjusted by the addition of safe and suitable pH-adjusting ingredients.

Whey Solids The solid fraction or dry form of whey. It is used as a replacement for milk solids—not-fat to provide a source of protein, solids, and flavor. It is used in baked goods, ice cream, dry mixes, and beverages.

Whole Fish Protein Concentrate A protein supplement that is derived from whole hake and hakelike fish, herring of the genera *Clupea,* menhaden, and anchovy of the species *Engraulis mordax.* The additive consists essentially of a dried fish protein processed from the whole fish without removal of heads, fins, tails, viscera, or intestinal contents. It is prepared by solvent extraction of fat and moisture with isopropyl alcohol or with ethylene dichloride followed by isopropyl alcohol, except that the additive derived from herring, menhaden, and anchovy is prepared by solvent extraction with isopropyl alcohol alone. Solvent residues are reduced by conventional heat drying and/or microwave radiation and there is a partial removal of bone.

Whole Milk See **Milk.**

Whole Milk Solids The product resulting from the drying or desiccation of milk. It contains not less than 26 percent fat and not more than 5 percent moisture. It is used in dry mixes such as puddings. It is also termed dried milk and milk powder.

Whole Wheat Flour The flour obtained by grinding cleaned wheat, other than durum wheat or red durum wheat, with the proportions of the natural constituents, other than moisture, remaining unaltered. The moisture content is not more than 15 percent. Optional ingredients include malted wheat, wheat flour, and barley flour for compensation for any natural deficiency of enzymes; ascorbic acid; and bleaching ingredients. It is also termed graham flour and entire wheat flour.

Wine Vinegar The vinegar made by the alcoholic and acetous fermentation of the juices of grapes or wine. It contains a minimum of 4 g per 100 cm^3 acid expressed as acetic acid. There is red wine vinegar, which has a rose to deep red color, and white wine vinegar, which has a pale yellow to off-white color. It is used in salad dressings, marinades, and sauces.

Worcestershire Sauce A sauce consisting of water, vinegar, soy sauce, corn syrup, salt and spices, or variations of these ingredients. It is used as a flavorant and is found in barbeque sauce and sweet-and-sour type sauces.

Xanthan Gum A gum obtained by microbial fermentation from the *Xanthomonas campestris* organism. It is very stable to viscosity change over varying temperatures, pH, and salt concentrations. It is also very pseudoplastic which results in a decrease in viscosity with increasing shear. It reacts synergistically with guar gum to provide an increase in viscosity and with carob gum to provide an increase in viscosity or gel formation. It is used in salad dressings, sauces, desserts, baked goods, and beverages at 0.05 to 0.50 percent.

Xylitol A sweetener that is a natural sugar substitute commercially made from the polysaccharide xylan obtained from birch trees. It is as sweet as sucrose and has a negative heat of solution which results in a cooling effect. It has a lower viscosity than sugar. It is used in chewing gum, throat lozenges, and chocolate.

Y

Yeast A leavening and fermentation agent that is a single-celled plant that can convert sugar to carbon dioxide. It is used as a leavening agent in bread and dough-type mixtures. It provides a yeasty flavor and tender crust. It has slow action as a leavening agent. One pound of active dry yeast replaces approximately 2 pounds of fresh yeast. Selected yeast strains are used in wine fermentation.

Yeast Extract A flavor contributor and flavor enhancer consisting of a combination of nucleic acids, peptides, polypeptides, amino acids, and other constituents. It is obtained from the yeast cells of *Saccharomyces cerevisiae,* formed in the brewing of beer. It is used to provide the same functions as monosodium glutamate, although not to the same extent. It is used as a partial substitute for meat extract and also functions with other flavor ingredients such as hydrolyzed vegetable proteins. It is used in soups, gravies, spreads, dressings, and meat products. Typical usage levels range from 0.1 to 0.5 percent.

Yeast Food A complete food used in doughs. It contains dough conditioner ingredients such as calcium salts, sulfates, and phosphates which strengthen the gluten. It also contains ammonium salts and phosphates which function as yeast nutrients. It is used in bread dough and in the fermentation of alcoholic beverages.

Yeast-Malt Sprout Extract A food enhancer produced by partial hydrolysis of yeast extract (derived from *Saccharomyces cereviseae, Saccharo-*

myces fragilis, or *Candida utilis)* using the sprout portion of malt barley as the source of enzymes. The additive contains a maximum of 6 percent 5′ nucleotides by weight.

Yellow Prussiate of Soda An anticaking agent and crystallizing agent. It is sometimes added as a crystallizing agent to salt when it crystallizes to generate jagged and bulky crystals which resist caking. It also functions as a water-soluble anticaking agent. It is also termed sodium ferrocyanide.

Yogurt A custard-like or soft gel product made by fermenting milk with bacterial cultures, specifically *Lactobacillus bulgaricus* and *Streptococcus thermophilus,* to a pH range of usually 4.0 to 4.5. It is used as a snack; as a meal; or in desserts, salad dressings, and baked goods.

Yucca Plant Extract A foaming agent obtained from the yucca plant species *Yucca brevifolia* and *Yucca schidigera.* It is available in liquid concentrate or dried form, is dark brown, and has a slight bittersweet flavor and a pH of 4.0. It is stable over a wide pH range and heat treatment and is readily soluble in water. It is used in applications where a frothy appearance and foam stability are desired, such as in root beer, cocktail mixes, and whipped beverages. Usage level is 50 to 150 parts per million.

Z

Zein A corn protein produced from corn gluten meal. It lacks the amino acids lysine and tryptophan, so it is not suitable as a sole source of dietary protein. It is insoluble in water and alcohols but is soluble in aqueous alcohols, glycols, and glycol ethers. It functions as a film and coating to provide a moisture barrier for nuts and grain products. It also functions as a coating for confections and a glaze for panned goods.

Zinc Zn. A metallic element that functions as a nutrient and dietary supplement. It is believed to be necessary for nucleic acid metabolism, protein synthesis, and cell growth. Sources of zinc include zinc acetate, carbonate, chloride, gluconate, stearate, and sulfate.

Zinc Acetate A source of zinc that functions as a nutrient and dietary supplement.

Zinc Carbonate A source of zinc that functions as a nutrient and dietary supplement.

Zinc Chloride A source of zinc that functions as a nutrient and dietary supplement.

Zinc Gluconate A source of zinc that functions as a nutrient and dietary supplement.

Zinc Oxide A source of zinc that functions as a nutrient and dietary supplement.

Zinc Methionine Sulfate A source of dietary zinc that is the product of the reaction between equimolar amounts of zinc sulfate and DL-methionine in purified water. It is used in tablet form.

Zinc Stearate A source of zinc that functions as a nutrient and dietary supplement.

Zinc Sulfate A source of zinc that functions as a nutrient and dietary supplement. It exists as prisms, needles, or powder. It has a solubility of 1 g in 0.6 ml of water. It is found in frozen egg substitutes.

Substances for Use in Foods

Listing Under Title 21 of the Code of Federal Regulations

Part 73. Listing of Color Additives Exempt From Certification

These substances may be safely used as diluents in color additive mixtures for food use exempt from certification, subject to the condition that each straight color in the mixture has been exempted from certification. If not exempted, the color is from a batch that has previously been certified and has not changed in composition since certification.

Subpart A—Foods

73.1 Diluents in color additive mixtures for food use exempt from certification
73.30 Annatto extract
73.40 Dehydrated beets (beet powder)
73.50 Ultramarine blue
73.75 Canthaxanthin
73.85 Caramel
73.90 β-Apo-8′-carotenal
73.95 β-Carotene
73.100 Cochineal extract; carmine
73.140 Toasted partially defatted cooked cottonseed flour
73.160 Ferrous gluconate
73.169 Grape color extract
73.170 Grape skin extract (enocianina)
73.200 Synthetic iron oxide
73.250 Fruit juice

73.260 Vegetable juice
73.275 Dried algae meal
73.295 Tagetes (Aztec marigold) meal and extract
73.300 Carrot oil
73.315 Corn endosperm oil
73.340 Paprika
73.345 Paprika oleoresin
73.450 Riboflavin
73.500 Saffron
73.575 Titanium dioxide
73.600 Turmeric
73.614 Turmeric oleoresin

Part 74. Listing of Color Additives Subject to Certification

These color additives may be safely used for coloring foods according to specified uses and restrictions.

Subpart A—Foods

74.101 FD&C Blue #1
74.102 FD&C Blue #2
74.203 FD&C Green #3
74.250 Orange B
74.302 Citrus Red #2
74.303 FD&C Red #3
74.340 FD&C Red #40
74.705 FD&C Yellow #5
74.706 FD&C Yellow #6

Part 172. Food Additives Permitted for Direct Addition to Food for Human Consumption

The listed food additives are permitted for direct addition to food for human consumption. Their usage is under conditions of good manufacturing practice.

Subpart A—General Provisions

172.5 General provisions for direct food additives

Subpart B—Food Preservatives

172.105 Anoxomer
172.110 BHA
172.115 BHT
172.120 Calcium disodium EDTA
172.130 Dehydroacetic acid
172.133 Dimethyl dicarbonate
172.135 Disodium EDTA
172.140 Ethoxyquin
172.145 Heptylparaben
172.150 4-Hydroxymethyl-2,6-di-*tert*-butyl-phenol
172.155 Natamycin (pimaricin)
172.160 Potassium nitrate
172.165 Quaternary ammonium chloride combination
172.170 Sodium nitrate
172.175 Sodium nitrite
172.177 Sodium nitrite used in processing smoked chub
172.180 Stannous chloride
172.185 TBHQ
172.190 THBP

Subpart C—Coatings, Films, and Related Substances

172.210 Coatings on fresh citrus fruit
172.215 Coumarone-indene resin
172.225 Methyl esters of fatty acids produced from edible fats and oils
172.230 Microcapsules for flavoring substances
172.235 Morpholine
172.250 Petroleum naphtha
172.255 Polyacrylamide
172.260 Oxidized polyethylene
172.270 Sulfated butyl oleate
172.275 Synthetic paraffin and succinic derivatives
172.280 Terpene resin

Subpart D—Special Dietary and Nutritional Additives

172.310 Aluminum nicotinate
172.315 Nicotinamide-ascorbic acid complex
172.320 Amino acids
172.325 Bakers yeast protein
172.330 Calcium pantothenate, calcium chloride double salt
172.335 D-Pantothenamide
172.340 Fish protein isolate

172.345 Folic acid (folacin)
172.350 Fumaric acid and salts of fumaric acids
172.365 Kelp
172.370 Iron-choline citrate complex
172.372 N-Acetyl-L-methionine
172.375 Potassium iodide
172.385 Whole fish protein concentrate
172.395 Xylitol
172.399 Zinc methionine sulfate

Subpart E—Anticaking Agents

172.410 Calcium silicate
172.430 Iron ammonium citrate
172.480 Silicon dioxide
172.490 Yellow prussiate of soda

Subpart F—Flavoring Agents and Related Substances

172.510 Natural flavoring substances and natural substances used in conjunction with flavors
172.515 Synthetic flavoring substances and adjuvants
172.520 Cocoa with dioctyl sodium sulfoccinate for manufacturing
172.530 Disodium guanylate
172.535 Disodium inosinate
172.540 DL-Alanine
172.560 Modified hop extract
172.575 Quinine
172.580 Safrole-free extract of sassafras
172.585 Sugar beet extract flavor base
172.590 Yeast-malt sprout extract

Subpart G—Gums, Chewing Gum Bases and Related Substances

172.610 Arabinogalactan
172.615 Chewing gum base
172.620 Carrageenan
172.623 Carrageenan with polysorbate 80
172.626 Salts of carrageenan
172.655 Furcelleran
172.660 Salts of furcelleran
172.665 Gellan gum
172.695 Xanthan gum

Subpart H—Other Specific Usage Additives

172.710 Adjuvants for pesticide use dilutions

172.715 Calcium lignosulfonate
172.720 Calcium lactobionate
172.725 Gibberellic acid and its potassium salt
172.730 Potassium bromate
172.735 Glycerol ester of wood rosin
172.755 Stearyl monoglyceridyl citrate
172.765 Succistearin (stearoyl propylene glycol hydrogen succinate)
172.770 Ethylene oxide polymer
172.775 Methacrylic acid-divinylbenzene copolymer

Subpart I—Multipurpose Additives

172.800 Acesulfame potassium
172.802 Acetone peroxides
172.804 Aspartame
172.806 Azodicarbonamide
172.808 Copolymer condensates of ethylene oxide and propylene oxide
172.810 Dioctyl sodium sulfosuccinate
172.811 Glyceryl tristearate
172.812 Glycine
172.814 Hydroxylated lecithin
172.816 Methyl glucoside-coconut oil ester
172.818 Oxystearin
172.820 Polyethylene glycol (mean molecular weight 200–9,500)
172.822 Sodium lauryl sulfate
172.824 Sodium mono- and dimethyl naphthalene sulfonates
172.826 Sodium stearyl fumarate
172.828 Acetylated monoglycerides
172.830 Succinylated monoglycerides
172.832 Monoglyceride citrate
172.834 Ethoxylated mono- and diglycerides
172.836 Polysorbate 60
172.838 Polysorbate 65
172.840 Polysorbate 80
172.841 Polydextrose
172.842 Sorbitan monostearate
172.844 Calcium stearoyl-2-lactylate
172.846 Sodium stearoyl-2-lactylate
172.848 Lactylic esters of fatty acids
172.850 Lactylated fatty acid esters of glycerol and propylene glycol
172.852 Glyceryl-lacto esters of fatty acids

172.854	Polyglycerol esters of fatty acids
172.856	Propylene glycol mono- and diesters of fats and fatty acids
172.858	Propylene glycol alginate
172.859	Sucrose fatty acid esters
172.860	Fatty acids
172.861	Cocoa butter substitute from coconut oil, palm kernel oil, or both oils
172.862	Oleic acid derived from tall oil fatty acids
172.863	Salts of fatty acids
172.864	Synthetic fatty alcohols
172.866	Synthetic glycerin produced by the hydrogenolysis of carbohydrates
172.868	Ethyl cellulose
172.870	Hydroxypropyl cellulose
172.872	Methyl ethyl cellulose
172.874	Hydroxypropyl methylcellulose
172.876	Castor oil
172.878	White mineral oil
172.880	Petrolatum
172.882	Synthetic isoparaffinic petroleum hydrocarbons
172.884	Odorless light petroleum hydrocarbons
172.886	Petroleum wax
172.888	Synthetic petroleum wax
172.890	Rice bran wax
172.892	Food starch-modified
172.894	Modified cottonseed products intended for human consumption
172.896	Dried yeasts
172.898	Bakers yeast glycan

Part 182. Substances Generally Recognized as Safe

The substances listed are generally recognized as safe (GRAS) for then intended use and purpose, when used in accordance with good manufacturing practices. Good Manufacturing Practice includes (a) use of a quantity that does not exceed the amount reasonably required to accomplish its intended function; (b) the reduced use to the extent reasonably possible of the quantity of a substance that becomes a component of a food as a result of its use in the manufacturing, processing, or packaging of food but is not intended to accomplish any physical or technical effect; and (c) use of substances of appropriate food grade and prepared and handled as food ingredients.

Subpart A—General Provisions

182.1 Substances that are generally recognized as safe
182.10 Spices and other natural seasonings and flavorings
182.20 Essential oils, oleoresins (solvent-free), and natural extractives (including distillates)
182.40 Natural extractives (solvent-free) used in conjunction with spices, seasonings, and flavorings
182.50 Certain other spices, seasonings, essential oils, oleoresins, and natural extracts
182.60 Synthetic flavoring substances and adjuvants
182.70 Substances migrating from cotton and cotton fabrics used in dry food packaging
182.90 Substances migrating to food from paper and paperboard products
182.99 Adjuvants for pesticide chemicals

See specific sections below:

§182.10 SPICES AND OTHER NATURAL SEASONINGS AND FLAVORINGS

Common Name	Botanical Name of Plant Source
Alfalfa herb and seed	*Medicago sativa* L.
All spice	*Pimenta officinalis* Lindl
Ambrette seed	*Hibiscus abelmoschus* L.
Angelica	*Angelica archangelica* L. or other spp. of *Angelica*
Angelica root	"
Angelica seed	"
Angostura (cusparia bark)	*Galipea officinalis* Hancock
Anise	*Pimpinella anisum* L.
Anise, star	*Illicium verum* Hook. f.
Balm (lemon balm)	*Melissa officinalis* L.
Basil, bush	*Ocimum minimum* L.
Basil, sweet	*Ocimum basilicum* L.
Bay	*Laurus nobilis* L.
Calendula	*Calendula officinalis* L.
Camomile (chamomile), English or Roman	*Anthemis nobilis* L.
Camomile (chamomile), German or Hungarian	*Matricaria chamomilla* L.
Capers	*Cappans spinosa* L.
Capsicum	*Capsicum frutescens* L. or *Capsicum annuum* L.
Caraway	*Carum carvi* L.
Caraway, black (black cumin)	*Nigella sativa* L.

§182.10 (*cont'd*)

Common Name	Botanical Name of Plant Source
Cardamom (cardamon)	*Elettaria cardamomum* Maton
Cassia, Chinese	*Cinnamomum cassia* Blume
Cassia, Padang or Batavia	*Cinnamomum burmanni* Blume
Cassia, Saigon	*Cinnamomum loureirii* Nees
Cayenne pepper	*Capsicum frutescens* L. or *Capsicum annuum* L.
Celery seed	*Apium graveolens* L.
Chervil	*Anthriscus cerefolium* (L). Hoffm.
Chives	*Allium schoenoprasum* L.
Cinnamon, Ceylon	*Cinnamonum zeylanicum* Nees
Cinnamon, Chinese	*Cinnamomum cassia* Blume
Cinnamon, Saigon	*Cinnamomum loureirii* Nees
Clary (clary sage)	*Salvia sclarea* L.
Clover	*Trifolium* spp.
Coriander	*Coriandrum sativum* L.
Cumin (cummin)	*Cuminum cyminum* L.
Cumin, black (black caraway)	*Nigella sativa* L.
Elder flowers	*Sambucus canadensis* L.
Fennel, common	*Foeniculum vulgare* Mill
Fennel, sweet (finocchio, Florence fennel)	*Foeniculum vulgare* Mill, var. duice (DC.) Alex
Fenugreek	*Trigonella foenum-graecum* L.
Galanga (galangal)	*Alpinia officinarum* Hance
Geranium	*Pelargonium* spp.
Ginger	*Zingiber officinale* Rosc.
Grains of paradise	*Amomum melegueta* Rosc.
Horehound (hoarhound)	*Marrubium vulgare* L.
Horseradish	*Armoracia lapathifolia* Gilib
Hyssop	*Hyssopus officinalis* L.
Lavender	*Lavandula officinalis* Chaux
Linden flowers	*Tilia* spp.
Mace	*Myristica fragrans* Houtt
Marigold, pot	*Calendula officinalis* L.
Marjoram, pot	*Marjorana onites* (L.) Benth
Marjoram, sweet	*Marjorana hortensis* Moench
Mustard, black or brown	*Brassica nigra* (L.) Koch
Mustard, brown	*Brassica juncea* (L.) Coss
Mustard, white or yellow	*Brassica hirta* Moench
Nutmeg	*Myristica fragans* Houtt
Oregano (oreganum, Mexican oregano, Mexican sage, origan)	*Lippia* spp.
Paprika	*Capsicum annuum* L.
Parsley	*Petroselinum crispum* (Mill.) Mansf.
Pepper, black	*Piper nigrum* L.
Pepper, cayenne	*Capsicum frutescens* L. or *Capsicum annuum* L.
Pepper, red	"
Pepper, white	*Piper nigrum* L.

§182.10 (*cont'd*)

Common Name	Botanical Name of Plant Source
Peppermint	*Mentha piperita* L.
Poppy seed	*Papayer somniferum* L.
Pot marigold	*Calendula officinalis* L.
Pot marjoram	*Marjorana onites* (L.) Benth
Rosemary	*Rosmarinus officinalis* L.
Saffron	*Crocus sativus* L.
Sage	*Salvia officinalis* L.
Sage, Greek	*Salvia trioba* L.
Savory, summer	*Satureia hortensis* L. (Satureja)
Savory, winter	*Satureia montana* L. (Satureja)
Sesame	*Sesamum indicum* L.
Spearmint	*Mentha spicata* L.
Star anise	*Illicium verum* Hook. f.
Tarragon	*Artemisia dracunculus* L.
Thyme	*Thymus vulgaris* L.
Thyme, wild or creeping	*Thymus serpyllum* L.
Turmeric	*Curcuma longa* L.
Vanilla	*Vanilla planifolia* Andr. or *Vanilla tahitensis* J.W. Moore
Zedoary	*Curcuma zedoaria* Rosc.

§182.20 ESSENTIAL OILS, OLEORESINS (SOLVENT-FREE), AND NATURAL EXTRACTIVES (INCLUDING DISTILLATES)

Common Name	Botanical Name of Plant Source
Alfalfa	*Medicago sativa* L.
Allspice	*Pimenta officinalis* Lindl
Almond, bitter (from prussic acid)	*Prunus amygdalus* Batsch, *Prunus armeniaca* L., or *Prunus persica* (L.) Batsch
Ambrette (seed)	*Hibiscus moschatus* Moench
Angelica root	*Angelica archangelica* L.
Angelica seed	"
Angelica stem	"
Angostura (cusparia bark)	*Galipea officinalis* Hancock
Anise	*Pimpinella anisum* L.
Asafetida	*Ferula assa-foetida* L. and related spp. of *Ferula*
Balm (lemon balm)	*Melissa officinalis* L.
Balsam of Peru	*Myroxylon pereirae Klotzsch*
Basil	*Ocimum basilicum* L.
Bay leaves	*Laurus nobilis* L.
Bay (myrcia oil)	*Pimenta racemosa* (Mill.) J.W. Moore
Bergamot (bergamot orange)	*Citrus aurantium* L. subsp. bergamia Wright et Arn

§182.20 *(cont'd)*

Common Name	Botanical Name of Plant Source
Bitter almond (free from prussic acid)	*Prunus amygdalus* Batsch, *Prunus armeniaca* L., or *Prunus persica* (L.) Batsch
Bois de rose	*Aniba rosaeodora* Ducke
Cacao	*Theobroma cacao* L.
Camomile (chamomile) flowers, Hungarian	*Matricaria chamomilla* L.
Camomile (chamomile) flowers, Roman or English	*Anthemis nobilis* L.
Cananga	*Cananga odorata* Hook. f. and Thoms
Capsicum	*Capsicum frutescens* L. and *Capsicum annuum* L.
Caraway	*Carum carvi* L.
Cardamom seed (cardamon)	*Elettaria cardamomum* Maton
Carob bean	*Ceratonia siliqua* L.
Carrot	*Daucus carota* L.
Cascarilla bark	*Croton eluteria* Benn
Cassia bark, Chinese	*Cinnamomum cassia* Blume
Cassia bark, Padang or Batavia	*Cinnamomum burmanni* Blume
Cassia bark, Saigon	*Cinnamomum loureirii* Nees
Celery seed	*Apium graveolens* L.
Cherry, wild, bark	*Prunus serotina* Ehrh.
Chervil	*Anthriscus cerefolium* (L.) Hoffm.
Chicory	*Cichorium intybus* L.
Cinnamon bark, Ceylon	*Cinnamomum zeylanicum* Nees
Cinnamon bark, Chinese	*Cinnamomum cassia* Blume
Cinnamon bark, Saigon	*Cinnamomum loureirii* Nees
Cinnamon leaf, Ceylon	*Cinnamomum zeylanicum* Nees
Cinnamon leaf, Chinese	*Cinnamomum cassia* Blume
Cinnamon leaf, Saigon	*Cinnamomum loureirii* Nees
Citronella	*Cymbopogon nardus* Rendle
Citrus peels	*Citrus* spp.
Clary (clary sage)	*Salvia sclarea* L.
Clover	*Trifolium* spp.
Coca (decocainized)	*Erythroxylum coca* Lam. and other spp. of *Erythroxylum*
Coffee	*Coffea* spp.
Cola nut	*Cola acuminata* Schott and Endl., and other spp. of *Cola*
Coriander	*Coriandrum sativum* L.
Cumin (cummin)	*Cuminum cyminum* L.
Curacao orange peel (orange, bitter peel)	*Citrus aurantium* L.
Cusparia bark	*Galipea officinalis* Hancock
Dandelion	*Taraxacum officinale* Weber and T. *laevigatum* DC.
Dandelion root	"
Dog grass (quackgrass, triticum)	*Agropyron repens* (L.) Beauv.

§182.20 (*cont'd*)

Common Name	Botanical Name of Plant Source
Elder flowers	*Sambucus canadensis* L. and S. *nigra* l.
Estragole (esdragol, esdragon, tarragon)	*Artemisia dracunculus* L.
Estragon (tarragon)	"
Fennel, sweet	*Foeniculum vulgare* Mill
Fenugreek	*Trigonella foenum-graecum* L.
Galanga (galangal)	*Alpinia officinarum* Hance
Geranium	*Pelargonium* spp.
Geranium, East Indian	*Cymbopogon martini* Stapf.
Geranium, rose	*Pelargonium graveolens* L'Her.
Ginger	*Zingiber officinale* Rosc.
Grapefruit	*Citrus paradisi* Macf.
Guava	*Psidium* spp.
Hickory bark	*Carya* spp.
Horehound (hoarhound)	*Marrubium vulgare* L.
Hops	*Humulus lupulus* L.
Horsemint	*Monarda punctata* L.
Hyssop	*Hyssopus officinalis* L.
Immortelle	*Helichrysum augustifolium* DC.
Jasmine	*Jasminum officinale* L. and other spp. of *Jasminum*
Juniper (berries)	*Juniperus communis* L.
Kola nut	*Cola acuminata* Schott and Endl., and other spp. of *Cola*
Laurel berries	*Laurus nobilis* L.
Laurel leaves	*Laurus* spp.
Lavender	*Lavandula officinalis* Chaix
Lavender, spike	*Lavandula latifolia* Vill
Lavandin	Hybrids between *Lavandula officinalis* Chaix and *Lavandula latifolin* Vill
Lemon	*Citrus limon* (L.) Burm. f.
Lemon balm (see balm)	
Lemon grass	*Cymbopogon citratus* DC. and *Cymbopogon lexuosus* Stapf.
Lemon peel	*Citrus limon* (L.) Burm. f.
Lime	*Citrus aurantifolia* Swingle
Linden flowers	*Tilia* spp.
Locust bean	*Ceratonia siliqua* L.
Lupulin	*Humulus lupulus* L.
Mace	*Myristica fragans* Houtt
Mandarin	*Citrus reticulata* Blanco
Marjoram, sweet	*Marjorana hortensis* Moench
Mate	*Ilex paraguariensis* St. Hil.
Melissa (see balm)	
Menthol	*Mentha* spp.
Menthyl acetate	"
Molasses (extract)	*Saccarum officinarium* L.
Mustard	*Brassica* spp.
Maringin	*Citrus paradisi* Macf.

§182.20 (*cont'd*)

Common Name	Botanical Name of Plant Source
Neroli, bigarade	*Citrus aurantium* L.
Nutmeg	*Myristica fragans* Houtt
Onion	*Allium cepa* L.
Orange, bitter, flowers	*Citrus aurantium* L.
Orange, bitter, peel	"
Orange leaf	*Citrus sinensis* (L.) Osbeck
Orange, sweet	"
Orange, sweet, flowers	"
Orange, sweet, peel	"
Origanum	*Origanum* spp.
Palmarosa	*Cymbopogon martini* Stapf.
Paprika	*Capsicum annuum* L.
Parsley	*Petroselinum crispum* (Mill.) Mansf.
Pepper, black	*Piper nigrum* L.
Pepper, white	"
Peppermint	*Mentha piperita* L.
Peruvian balsam	*Myroxylon pereirae* Klotzsch
Petitgrain	*Citrus aurantium* L.
Petitgrain lemon	*Citrus limon* (L.) Burm. f.
Petitgrain mandarin or tangarine	*Citrus reticulata* Blanco
Pimenta	*Pimenta officinalis* Lindl
Pimenta leaf	"
Pipsissewa leaves	*Chimaphila umbellata* Nutt
Pomegranate	*Punica granatum* L.
Prickly ash bark	*Xanthoxylum* (or *Zanthoxylum Americanum* Mill. or *Xanthoxylum clava-herculis* L.
Rose absolute	*Rosa alba* L., *Rosa centifola* L., *Rosa damascena* Mill., *Rosa gallica* L., and vars. of these spp.
Rose (otto of roses, attar of roses)	"
Rose buds	"
Rose flowers	"
Rose fruit (hips)	"
Rose geranium	*Pelargonium graveolens* L'Her.
Rose leaves	*Rosa* spp.
Rosemary	*Rosmarinus officinalis* L.
Saffron	*Crocus sativus* L.
Sage	*Salvia officinalis* L.
Sage, Greek	*Salivia triloba* L.
Sage, Spanish	*Salvia lavandulaefolia* Vahl
St. John's bread	*Ceratonia siliqua* L.
Savory, summer	*Satureia hortensis* L.
Savory, winter	*Satureia montana* L.
Schinus molle	*Schinus molle* L.
Sloe berries (blackthorn berries)	*Prunus spinosa* L.
Spearmint	*Mentha spicata* L.
Spike lavender	*Lavandula latifolia* Vill.
Tamarind	*Tamarindus indica* L.

§182.20 (*cont'd*)

Common Name	Botanical Name of Plant Source
Tangerine	*Citrus reticulata* Blanco
Tarragon	*Artemisa dracunculus* L.
Tea	*Thea sinensis* L.
Thyme	*Thymus vulgaris* L. and *Thymus zygis* var. *gracilis* Boiss
Thyme, white	"
Thyme, wild or creeping	*Thymus serpyllum* L.
Triticum (see dog grass)	
Tuberose	*Polianthes tuberosa* L.
Tumeric	*Curcuma longa* L.
Vanilla	*Vanilla planifolia* Andr. or *Vanilla tahitensis* J.W. Moore
Violet flowers	*Viola odorata* L.
Violet leaves	"
Violet leaves absolute	"
Wild cherry bark	*Prunus serotina* Ehrh.
Ylang-ylang	*Cananga odorata* Hook. f. . and Thoms
Zedoary bark	*Curcuma zedonaria* Rosc.

§182.40 NATURAL EXTRACTIVES (SOLVENT-FREE) USED IN CONJUNCTION WITH SPICES, SEASONINGS, AND FLAVORINGS

Common Name	Botanical Name of Plant Source
Apricot kernel (persic oil)	*Prunus armeniaca* L.
Peach kernel (persic oil)	*Prunus persica* Sieb. et Zucc.
Peanut stearine	*Arachis hypogaea* L.
Persic oil (see apricot kernel and peach kernel)	
Quince seed	*Cydonia oblonga* Miller

§182.50 CERTAIN OTHER SPICES, SEASONINGS, ESSENTIAL OILS, OLEORESINS, AND NATURAL EXTRACTS

Common Name	Derivation
Ambergris	*Physeter macrocephalus* L.
Castoreum	*Castor fiber* L. and *C. canadensis* Kuhl
Civet (zibeth, zibet, zibetum)	Civet cats, *Viverra civetta* Schreber and *Viverra zibetha* Schreber
Cognac oil, white and green	Ethyl oenanthate, so-called
Musk (Tonquin musk)	Musk deer, *Moschus moschiferus* L.

§182.60 SYNTHETIC FLAVORING SUBSTANCES AND ADJUVANTS

Acetaldehyde (ethanol)
Acetoin (acetyl methylcarbinol)
Anethole (parapropenyl anisole)
Benzaldehyde (benzoic aldehyde)
N-Butyric acid (butanoic acid)
D- or L-Carvone (carvol)
Cinnamaldehyde (cinnamic aldehyde)
Citral (2,6-dimethyloctadien-2,6-*al*-8, geranial, neral)
Decanal (*N*-decylaldehyde, capraldehyde, capric aldehyde, caprinaldehyde, al dehyde C-10)
Ethyl acetate
Ethyl butyrate
3-Methyl-3-phenyl glycidic acid ethyl ester (ethyl-menthyl-phenyl-glycidate, so called strawberry aldehyde, C-16 aldehyde)
Ethyl vanillin
Geraniol (3,7-dimethyl-2,6 and 3,6-octadien-1-*ol*)
Geranyl acetate (geraniol acetate)
Limonene (D-, L-, and DL-)
Linalool (linalol, 3,7-dimethyl-1,6-octadien-3-*ol*)
Linalyl acetate (bergamol)
Methyl anthranilate (methyl-2-aminobenzoate)
Piperonal (3,4-methylenedioxy-benzaldehyde, heliotropin)
Vanillin

§182.70 SUBSTANCES MIGRATING FROM COTTON AND COTTON FABRICS USED IN DRY FOOD PACKAGING

Beef tallow
Carboxymethylcellulose
Coconut oil, refined
Cornstarch
Gelatin
Japan wax
Lard
Lard oil
Oleic acid
Peanut oil
Potato starch
Sodium acetate
Sodium chloride
Sodium silicate
Sodium tripolyphosphate
Soybean oil (hydrogenated)
Talc
Tallow (hydrogenated)
Tallow flakes
Tapioca starch
Tetrasodium pyrophosphate

Wheat starch
Zinc chloride

---

§182.90 SUBSTANCES MIGRATING TO FOOD FROM PAPER AND PAPERBOARD PRODUCTS

Alum (double sulfate of aluminum and ammonium potassium, or sodium)
Aluminum hydroxide
Aluminum oleate
Aluminum palmitate
Casein
Cellulose acetate
Cornstarch
Diatomaceous earth filler
Ethyl cellulose
Ethyl vanillin
Glycerin
Mono- and diglycerides from glycerolysis of edible fats and oils
Oleic acid
Potassium sorbate
Silicon dioxides
Sodium aluminate
Sodium chloride
Sodium hexametaphosphate
Sodium hydrosulfite
Sodium phosphoaluminate
Sodium silicate
Sodium sorbate
Sodium tripolyphosphate
Sorbitol
Soy protein, isolated
Starch, acid-modified
Starch, pregelatinized
Starch, unmodified
Talc
Vanillin
Zinc hydrosulfite
Zinc sulfate

Subpart B—Multiple Purpose GRAS Food Substances

182.1033 Citric acid
182.1045 Glutamic acid
182.1047 Glutamic acid hydrochloride
182.1057 Hydrochloric acid
182.1073 Phosphoric acid
182.1087 Sodium acid pyrophosphate
182.1125 Aluminum sulfate

182.1127 Aluminum ammonium sulfate
182.1129 Aluminum potassium sulfate
182.1131 Aluminum sodium sulfate
182.1180 Caffeine
182.1195 Calcium citrate
182.1217 Calcium phosphate
182.1235 Caramel
182.1320 Glycerin
182.1324 Glyceryl monostearate
182.1480 Methylcellulose
182.1500 Monoammonium glutamate
182.1516 Monopotassium glutamate
182.1625 Potassium citrate
182.1711 Silica aerogel
182.1745 Sodium carboxymethylcellulose
182.1748 Sodium caseinate
182.1751 Sodium citrate
182.1778 Sodium phosphate
182.1781 Sodium aluminum phosphate
182.1810 Sodium tripolyphosphate
182.1866 High-fructose corn syrup
182.1911 Triethyl citrate

Subpart C—Anticaking Agents

182.2122 Aluminum calcium silicate
182.2227 Calcium silicate
182.2437 Magnesium silicate
182.2727 Sodium aluminosilicate
182.2729 Sodium calcium aluminosilicate, hydrated
182.2906 Tricalcium silicate

Subpart D—Chemical Preservatives

182.3013 Ascorbic acid
182.3041 Erythorbic acid
182.3089 Sorbic acid
182.3109 Thiodipropionic acid
182.3149 Ascorbyl palmitate
182.3169 Butylated hydroxyanisole
182.3173 Butylated hydroxytoluene
182.3189 Calcium ascorbate
182.3225 Calcium sorbate
182.3280 Dilauryl thiodipropionate
182.3616 Potassium bisulfite

182.3637 Potassium metabisulfite
182.3640 Potassium sorbate
182.3731 Sodium ascorbate
182.3739 Sodium bisulfite
182.3766 Sodium metabisulfite
182.3795 Sodium sorbate
182.3798 Sodium sulfite
182.3862 Sulfur dioxide
182.3890 Tocopherols

Subpart E—Emulsifying Agents

Subpart F—Dietary Supplements

182.5013 Ascorbic acid
182.5065 Linoleic acid
182.5159 Biotin
182.5191 Calcium carbonate
182.5195 Calcium citrate
182.5201 Calcium glycerophosphate
182.5210 Calcium oxide
182.5212 Calcium pantothenate
182.5217 Calcium phosphate
182.5223 Calcium pyrophosphate
182.5245 Carotene
182.5250 Choline bitartrate
182.5252 Choline chloride
182.5260 Copper gluconate
182.5301 Ferric phosphate
182.5304 Ferric pyrophosphate
182.5306 Ferric sodium pyrophosphate
182.5308 Ferrous gluconate
182.5311 Ferrous lactate
182.5315 Ferrous sulfate
182.5370 Inositol
182.5375 Iron reduced
182.5431 Magnesium oxide
182.5434 Magnesium phosphate
182.5443 Magnesium sulfate
182.5446 Manganese chloride
182.5449 Manganese citrate
182.5452 Manganese gluconate
182.5455 Manganese glycerophosphate
182.5461 Manganese sulfate

182.5464	Manganous oxide
182.5530	Niacin
182.5535	Niacinamide
182.5580	D-Pantothenyl alcohol
182.5622	Potassium chloride
182.5628	Potassium glycerophosphate
182.5676	Pyridoxine hydrochloride
182.5695	Riboflavin
182.5697	Riboflavin-5-phosphate
182.5772	Sodium pantothenate
182.5778	Sodium phosphate
182.5875	Thiamine hydrochloride
182.5878	Thiamine mononitrate
182.5890	Tocopherols
182.5892	α-Tocopherol acetate
182.5930	Vitamin A
182.5933	Vitamin A acetate
182.5936	Vitamin A palmitate
182.5945	Vitamin B_{12}
182.5950	Vitamin D_2
182.5953	Vitamin D_3
182.5985	Zinc chloride
182.5988	Zinc gluconate
182.5991	Zinc oxide
188.5994	Zinc stearate
182.5997	Zinc sulfate

Subpart G—Sequestrants

182.6033	Citric acid
182.6085	Sodium acid phosphate
182.6195	Calcium citrate
182.6197	Calcium diacetate
182.6203	Calcium hexametaphosphate
182.6215	Monobasic calcium phosphate
182.6285	Dipotassium phosphate
182.6290	Disodium phosphate
182.6386	Isopropyl citrate
182.6511	Monoisopropyl citrate
182.6625	Potassium citrate
182.6751	Sodium citrate
182.6757	Sodium gluconate
182.6760	Sodium hexametaphosphate
182.6769	Sodium metaphosphate

182.6778 Sodium phosphate
182.6787 Sodium pyrophosphate
182.6789 Tetrasodium pyrophosphate
182.6810 Sodium tripolyphosphate
182.6851 Stearyl citrate

Subpart H—Stabilizers
182.7255 Chondrus extract

Subpart I—Nutrients
182.8013 Ascorbic acid
182.8159 Biotin
182.8195 Calcium citrate
182.8217 Calcium phosphate
182.8223 Calcium pyrophosphate
182.8250 Choline bitartrate
182.8252 Choline chloride
182.8458 Manganese hypophosphite
182.8778 Sodium phosphate
182.8890 Tocopherols
182.8892 α-Tocopherol acetate
182.8985 Zinc chloride
182.8988 Zinc gluconate
182.8991 Zionc oxide
182.8994 Zinc stearate
182.8997 Zinc sulfate

Part 184. Direct Food Substances Affirmed As Generally Recognized As Safe

The substances listed have been reviewed and determined by the Food and Drug Administration to be generally recognized as safe (GRAS) for the purposes and the other conditions prescribed. They shall be used in accordance with current good manufacturing practice.

Subpart A—General Provisions
184.1 Substances added directly to human food affirmed as generally recognized as safe (GRAS)

Subpart B—Listing of Specific Substances Affirmed as GRAS
184.1005 Acetic acid
184.1007 Aconitic acid

184.1009 Adipic acid
184.1011 Alginic acid
184.1021 Benzoic acid
184.1025 Caprylic acid
184.1027 Mixed carbohydrase and protease enzyme product
184.1061 Lactic acid
184.1065 Linoleic acid
184.1069 Malic acid
184.1077 Potassium acid tartrate
184.1081 Propionic acid
184.1090 Stearic acid
184.1091 Succinic acid
184.1095 Sulfuric acid
184.1097 Tannic acid
184.1099 Tartaric acid
184.1101 Diacetyl tartaric acid esters of mono- and diglycerides
184.1115 Agar-agar
184.1120 Brown algae
184.1121 Red algae
184.1133 Ammonium alginate
184.1135 Ammonium bicarbonate
184.1137 Ammonium carbonate
184.1138 Ammonium chloride
184.1139 Ammonium hydroxide
184.1141a Ammonium phosphate, monobasic
184.1141b Ammonium phosphate, dibasic
184.1143 Ammonium sulfate
184.1155 Bentonite
184.1157 Benzoyl peroxide
184.1165 *n*-Butane and isobutane
184.1185 Calcium acetate
184.1187 Calcium alginate
184.1191 Calcium carbonate
184.1193 Calcium chloride
184.1199 Calcium gluconate
184.1201 Calcium glycerophosphate
184.1205 Calcium hydroxide
184.1206 Calcium iodate
184.1207 Calcium lactate
184.1210 Calcium oxide
184.1212 Calcium pantothenate
184.1221 Calcium propionate
184.1229 Calcium stearate

184.1230	Calcium sulfate
184.1240	Carbon dioxide
184.1245	Beta-carotene
184.1257	Clove and its derivatives
184.1259	Cocoa butter substitute primarily from palm oil
184.1260	Copper gluconate
184.1261	Copper sulfate
184.1262	Corn silk and corn silk extract
184.1265	Cuprous iodide
184.1271	L-Cysteine
184.1272	L-Cysteine monohydrochloride
184.1277	Dextrin
184.1278	Diacetyl
184.1282	Dill and its derivatives
184.1287	Enzyme-modified fats
184.1293	Ethyl alcohol
184.1295	Ethyl formate
184.1296	Ferric ammonium citrate
184.1297	Ferric chloride
184.1298	Ferric citrate
184.1301	Ferric phosphate
184.1304	Ferric pyrophosphate
184.1307	Ferric sulfate
184.1307a	Ferrous ascorbate
184.1307b	Ferrous carbonate
184.1307c	Ferrous citrate
184.1307d	Ferrous fumarate
184.1308	Ferrous gluconate
184.1311	Ferrous lactate
184.1315	Ferrous sulfate
184.1317	Garlic and its derivatives
184.1318	Glucono delta-lactone
184.1321	Corn gluten
184.1322	Wheat gluten
184.1323	Glyceryl monooleate
184.1324	Glyceryl monostearate
184.1328	Glyceryl behenate
184.1330	Acacia (gum arabic)
184.1333	Gum ghatti
184.1339	Guar gum
184.1343	Locust (carob) bean gum
184.1349	Karaya gum (sterculia gum)
184.1351	Gum tragacanth

184.1355	Helium
184.1366	Hydrogen peroxide
184.1370	Inositol
184.1372	Insoluble glucose isomerase enzyme preparations
184.1375	Iron, elemental
184.1388	Lactase enzyme preparation from *Kluyveromye lactis*
184.1400	Lecithin
184.1408	Licorice and licorice derivatives
184.1409	Ground limestone
184.1425	Magnesium carbonate
184.1426	Magnesium chloride
184.1428	Magnesium hydroxide
184.1431	Magnesium oxide
184.1434	Magnesium phosphate
184.1440	Magnesium stearate
184.1443	Magnesium sulfate
184.1444	Maltodextrin
184.1445	Malt syrup (malt extract)
184.1446	Manganese chloride
184.1449	Manganese citrate
184.1452	Manganese gluconate
184.1461	Manganese sulfate
184.1472	Hydrogenated and partially hydrogenated menhaden oils
184.1490	Methylparaben
184.1498	Microparticulated protein product
184.1505	Mono- and diglycerides
184.1521	Monosodium phosphate derivatives of mono- and diglycerides
184.1530	Niacin
184.1535	Niacinamide
184.1537	Nickel
184.1538	Nisin preparation
184.1540	Nitrogen
184.1545	Nitrous oxide
184.1553	Peptones
184.1555	Rapeseed oil
184.1560	Ox bile extract
184.1563	Ozone
184.1585	Papain
184.1588	Pectins
184.1610	Potassium alginate
184.1613	Potassium bicarbonate
184.1619	Potassium carbonate

184.1622 Potassium chloride
184.1631 Potassium hydroxide
184.1634 Potassium iodide
184.1635 Potassium iodate
184.1639 Potassium lactate
184.1643 Potassium sulfate
184.1655 Propane
184.1660 Propyl gallate
184.1666 Propylene glycol
184.1670 Propylparaben
184.1676 Pyridoxine hydrochloride
184.1685 Rennet (animal-derived)
184.1695 Riboflavin
184.1697 Riboflavin-5′-phosphate (sodium)
184.1698 Rue
184.1699 Oil of rue
184.1721 Sodium acetate
184.1724 Sodium alginate
184.1733 Sodium benzoate
184.1736 Sodium bicarbonate
184.1742 Sodium carbonate
184.1754 Sodium diacetate
184.1763 Sodium hydroxide
184.1764 Sodium hypophosphite
184.1768 Sodium lactate
184.1769a Sodium metasilicate
184.1784 Sodium propionate
184.1792 Sodium sesquicarbonate
184.1801 Sodium tartrate
184.1804 Sodium potassium tartrate
184.1807 Sodium thiosulfate
184.1835 Sorbitol
184.1845 Stannous chloride (anhydrous and dihydrated)
184.1848 Starter distillate
184.1854 Sucrose
184.1857 Corn sugar
184.1859 Invert sugar
184.1865 Corn syrup
184.1875 Thiamine hydrochloride
184.1878 Thiamine mononitrate
184.1890 α-Tocopherols
184.1901 Triacetin
184.1903 Tributyrin

184.1923 Urea
184.1924 Urease enzyme preparation from *Lactobacillus fermentum*
184.1930 Vitamin A
184.1945 Vitamin B_{12}
184.1950 Vitamin D
184.1973 Beeswax (yellow and white)
184.1976 Candelilla wax
184.1978 Carnauba wax
184.1979 Whey
184.1979a Reduced lactose whey
184.1979b Reduced minerals whey
184.1979c Whey protein concentrate
184.1983 Bakers yeast extract
184.1984 Zein

Part 186. Indirect Food Substances Affirmed As Generally Recognized As Safe

These substances are added indirectly to human food for the purpose and under the conditions prescribed, providing the substances comply with the listed purity specifications or are of a purity suitable for their intended use. These substances have been reviewed and are generally recognized as safe.

Subpart A—General Provisions

186.1 Substances added indirectly to human food affirmed as generally recognized as safe (GRAS)

Subpart B—Listing of Specific Substances Affirmed as GRAS

186.1025 Caprylic acid
186.1093 Sulfamic acid
186.1256 Clay (kaolin)
186.1275 Dextrans
186.1300 Ferric oxide
186.1316 Formic acid
186.1374 Iron oxides
186.1551 Hydrogenated fish oil
186.1557 Tall oil
186.1673 Pulp
186.1750 Sodium chlorite
186.1756 Sodium formate

186.1770	Sodium oleate
186.1771	Sodium palmitate
186.1797	Sodium sulfate
186.1839	Sorbose

Bibliography

Acree, T. E., and R. Teranishi. 1993. *Flavor Science: Sensible Principles and Techniques.* Washington, DC: ACS.

Akoh, C. C., and B. G. Swanson. 1994. *Carbohydrate Polyesters as Fat Substitutes.* New York: Marcel Dekker, Inc.

Alais, C., and G. Linden. 1991. *Food Biochemistry.* New York: Chapman and Hall.

Almond, N. 1989. *Biscuits, Cookies and Crackers.* Volume 2. New York: Chapman and Hall.

Almond, N., et al. 1991. *Biscuits, Cookies and Crackers.* Volume 3. New York: Chapman and Hall.

Altschul, A. 1993. *Low-Calorie Foods Handbook.* New York: Marcel Dekker.

Anonymous. 1988. *Canned Foods: Principles of Thermal Process Control, Acidification and Container Closure Evaluation.* 5th ed. Washington, DC: Food Processors Institute.

Anonymous. 1988. *Industrial Egg Handbook.* Upland, CA: California Egg Commission.

Anonymous. 1988. *New Ideas: The Pork Technical Reference Manual.* Chicago: Pork Industry Group, National Live Stock & Meat Board.

Anonymous. 1989. *Leading Edge Reports: Synthetic and Organic Food Additives Markets.* Cleveland Heights, OH: Leading Edge Group.

Anonymous. 1989. *Recommended Dietary Allowances.* 10th ed. Washington, DC: National Academic Press.

Anonymous. 1989. *Requirements of Laws and Regulations Enforced by the U.S. Food and Drug Administration.* Washington, DC: Government Printing Office.

Anonymous. 1990. *Leading Edge Reports: A Competitive Survey of Flavor and Fragrance Markets.* Cleveland Heights, OH: Leading Edge Group.

Anonymous. 1991. *Flavor and Fragrance Materials—1991.* Wheaton, IL: Allured Publishing Corporation.
Anonymous. 1991. *Food Fats and Health.* Ames, IA: Council for Agricultural Science and Technology.
Anonymous. 1991. *Nutritional and Toxiocological Consequences of Food Processing.* New York: Plenum Publishing Corp.
Anonymous. 1992. *Maize in Human Nutrition.* Lanham, MD: Food and Agriculture Organization of the United Nations.
Anonymous. 1993. *The Almanac of the Canning, Freezing, Preserving Industries.* 7th ed. Volume 1. Westminster, MD: Edward E. Judge & Sons.
Anonymous. 1993. *The Almanac of the Canning, Freezing, Preserving Industries.* 7th ed. Volume 2. Westminster, MD: Edward E. Judge & Sons.
Anonymous. 1993. *Diet and Cancer.* 2nd ed. New York: American Council on Science and Health Inc.
Anonymous. 1993. *Everything Added to Foods in the United States.* Boca Raton, FL: CRC Press.
Anonymous. 1993. *Flavor and Fragrance Materials 1993.* Carol Stream, IL: Allured Publishing.
Arbuckle, W. S. 1986. *Ice Cream.* 4th ed. New York: Chapman and Hall.
Aruoma, O. I. and B. Halliwell. 1991. *Free Radicals and Food Additives.* Bristol, PA: Taylor & Francis Group.
Ashurst, P. R. 1991. *Food Flavourings.* New York: Chapman and Hall.
Baianu, I. C. 1992. *Physical Chemistry of Food Processes.* Volume 1. New York: Chapman and Hall.
Baianu, I. C., et al. 1993. *Physical Chemistry of Food Processes.* Volume II. New York: Chapman and Hall.
Bald, W. B. 1991. *Food Freezing: Today and Tomorrow.* Secaucus, NJ: Springer-Verlag New York.
Barrows, A. B. 1985. *Bakery Specialities.* New York: Chapman and Hall.
Bauernfeind, J. C., and P. A. Lachance. 1991. *Nutrient Additions to Food: Nutritional, Technological and Regulatory Aspects.* Trumbull, CT: Food & Nutrition Press.
Beckett, S. 1994. *Industrial Chocolate Manufacture and Use.* 2nd ed. New York: Chapman and Hall.
Bee, R. D., et al. 1989. *Food Colloids.* Boca Raton, FL: CRC Press.
Bender, A. E. 1990. *Dictionary of Nutrition and Food Technology.* Stoneham, MA: Butterworths.
Bennion, E. B., and G. S. T. Bamford. 1986. *The Technology of Cake Making.* New York: Chapman and Hall.
Bessiere, Y., and A. F. Thomas. 1990. *Flavour Science and Technology.* New York: John Wiley & Sons.
Binkley, R. W. 1988. *Modern Carbohydrate Chemistry.* New York: Marcel Dekker.

Birch, G. G., and K. J. Parker. 1979. *Sugar: Science and Technology.* New York: Chapman and Hall.
Birch, G. G., et al. 1971. *Sweetness and Sweeteners.* New York: Chapman and Hall.
Bishop, R. J. 1994. *Dairy Technology Handbook.* New York: Chapman and Hall.
Block, S. S. 1991. *Disinfection, Sterilization, and Preservation.* 4th ed. York, PA: Williams & Wilkins.
Bonnell, A. D. 1994. *Quality Assurance in Seafood Processing: A Practical Guide.* New York: Chapman & Hall.
Booth, R. G. 1991. *Snack Food.* New York: Chapman and Hall.
Branen, A. L., et al. *Food Additives.* New York: Marcel Dekker.
Brody, T. 1994. *Nutritional Biochemistry.* San Diego: Academic Press.
Brown, M. L. 1990. *Present Knowledge in Nutrition.* 6th ed. Washington, DC: International Life Sciences Institute—Nutrition Foundation.
Buckingham, J. 1994. *Dictionary of Natural Products. The Main Work (in seven volumes).* New York: Chapman and Hall.
Burke, T. A. et al. *Regulating Risk, The Science and Politics of Risk.* Washington, DC: ILSI Press.
Bushuk, W., and V. F. Rasper. 1994. *Wheat: Production, Properties and Quality.* New York: Chapman & Hall.
Bushuk, W., and Tkachuk, R. 1991. *Gluten Proteins 1990.* St. Paul: American Assn. of Cereal Chemists.
Cambie, R. C. 1989. *Fats for the Future.* New York: Chapman and Hall.
Cassens, R. G. 1990. *Nitrite-Cured Meat: A Food Safety Issue in Perspective.* Trumbull, CT: Food and Nutrition Press.
Catsberg, C. M. E., and G. J. M. Kempen-van Dommelen. 1990. *Food Handbook.* New York: Chapman and Hall.
Chalmers, I. 1990. *The Food Professional's Guide.* New York: American Showcase.
Chandan, R. C., et al. *Yogurt: Nutritional Health Properties.* McLean, VA: National Yogurt Association.
Charalambous, G. 1988. *Frontiers of Flavor.* New York: Elsevier Science Publishing.
Charalambous, G. 1990. *Flavours and Off-Flavours '89.* New York: Elsevier Applied Science Publishing.
Charalambous, G. 1992. *Food Science and Human Nutrition.* New York: Elsevier Science Publishing.
Charalambous, G. 1993. *Food Flavors, Ingredients and Composition.* New York: Elsevier Science Publishing.
Charalambous, G., and G. Doxastakis. 1989. *Food Emulsifiers: Chemistry, Technology, Functional Properties and Applications.* New York: Elsevier Science Publishing.

Charambous, G. 1992. *Off-Flavors in Foods and Beverages.* New York: Elsevier Science Publishing.

Combs, Jr., G. F. 1991. *The Vitamins: Fundamental Aspects in Nutrition and Health.* San Diego: Academic Press.

Considine, M. D., and G. D. Considine. 1982. *Foods and Food Production Encyclopedia.* New York: Chapman and Hall.

Craker, L. E., and J. E. Simon. 1988. *Herbs, Spices and Medicinal Plants: Recent Advances in Botany, Horticulture, and Pharmacology,* Volume 3. Phoenix: Oryx Press.

Daniel, A. R. 1971. *The Bakers' Dictionary.* 2nd ed. New York: Chapman and Hall.

Davidson, M., and Branen, A. L. 1993. *Antimicrobials in Foods.* 2nd ed. New York: Marcel Dekker.

de Man, J. M. 1989. *Principles of Food Chemistry.* 2nd ed. New York: Van Nostrand Reinhold.

Dickinson, E. 1992. *An Introduction to Food Colloids.* Cary, NC: Oxford University Press.

Dickinson, E., and G. Stainsby. 1988. *Advances in Food Emulsions and Foams.* New York: Chapman and Hall.

Dobbin, J. 1989. *Dietary Starches and Sugars in Man: A Comparison.* New York: Springer-Verlag.

Eck, P. 1990. *The American Cranberry.* New Brunswick, NJ: Rutgers University Press.

Ecobichon, D. J. *The Basis of Toxicity Testing.* 1992. Boca Raton, FL: CRC Press.

Egan, M., and S. Davis Allen. 1992. *Healthful Quantity Baking.* New York: John Wiley & Sons.

El-Nokaly, M., and D. Cornell. 1991. *Microemulsions and Emulsions in Food.* York, PA: American Chemical Society.

Eliasson, A., and Larsson, K. 1993. *Cereals in Breadmaking: A Molecular Colloidal Approach.* New York: Marcel Dekker.

Ensminger, A. H., et al. 1993. *Foods & Nutrition Encyclopedia.* 2nd ed. Volume I—A–H; Volume II—I–Z. Boca Raton, FL: CRC Press.

Eskin, N. A. M. 1990. *Biochemistry of Foods.* 2nd ed. San Diego, CA: Academic Press.

Faridi, H. 1994. *The Science of Cookie and Cracker Production.* New York: Chapman & Hall.

Farrell, K. T. 1991. *Spices, Condiments, and Seasonings.* 2nd ed. New York: Chapman and Hall.

Fennema, O. R. 1975. *Principles of Food Science.* (Food Science and Technology Series, Volume 4). New York: Marcel Dekker.

Fennema, O. R. 1985. *Food Chemistry: Revised and Expanded.* (Food Science and Technology Series, Volume 15). 2nd ed. New York: Marcel Dekker.

Fennema, O. R., et al. 1973. *Low-Temperature Preservation of Foods and Living Matter.* (Food Science and Technology Series, Volume 3). New York: Marcel Dekker.

Finley, H. W., et al. 1992. *Food Safety Assessment.* Washington, DC: American Chemical Society.

Flick, E. W. 1990. *Emulsifying Agents: An Industrial Guide.* Park Ridge, NJ: Noyes Publications.

Footitt, R., and A. S. Lewis. 1994. *Technology of Meat and Fish Canning.* New York: Chapman and Hall.

Fox, P. F. 1983. *Developments in Dairy Chemistry.* Volume 2. New York: Chapman and Hall.

Fox, P. F. 1985. *Developments in Dairy Chemistry.* Volume 3. New York: Chapman and Hall.

Fox, P. F. 1989. *Developments in Dairy Chemistry.* Volume 4. New York: Chapman and Hall.

Fox, P. F. 1993. *Advanced Dairy Chemistry 1: Proteins.* New York: Chapman and Hall.

Fox, P. F. 1993. *Cheese: Chemistry, Physics and Microbiology.* Volume 1. 2nd ed. New York: Chapman and Hall.

Fox, P. F. 1993. *Cheese: Chemistry, Physics and Microbiology.* Volume 2. 2nd ed. New York: Chapman and Hall.

Fox, P. F. and J. J. Condon. 1982. *Food Proteins.* New York: Chapman and Hall.

Frank, R. C., and H. B. Irving. 1992. *The Directory of Food and Nutrition Information for Professionals and Consumers.* 2nd ed. Phoenix: The Orynx Press.

Freudenthal, R. I., and S. L. Freudenthal. 1991. *Food: Facts and Fiction.* Boca Raton, FL: Hill and Garnett Publishing.

Friedman, M. 1989. *Absorption and Utilization of Amino Acids.* Volumes I, II, and III. Boca Raton, FL: CRC Press.

Fuller, D. B., and R. T. Parry. 1986. *Savory Coatings.* New York: Chapman and Hall.

Fuller, G. W. 1994. *New Food Product Development: From Concept to Marketplace.* Boca Raton, FL: CRC Press.

Gaby, S. K., et al. 1991. *Vitamin Intake and Health: A Scientific Review.* New York: Marcel Dekker.

Gallagher, C. R., and J. B. Allred. 1992. *Taking the Fear Out of Eating: A Nutritionist's Guide to Sensible Food Choices.* New York: Cambridge University Press.

Gaull, G. E., and Goldberg, R. A. 1991. *New Technologies and the Future of Food and Nutrition.* New York: John Wiley & Sons.

Gordon, B., and C. William. 1994. *Primary Cereal Processing: A Comprehensive Sourcebook.* New York: VCH Publishers.

Goldberg, I. 1994. *Functional Foods: Designer Foods, Pharmafoods, Nutraceuticals.* New York: Chapman & Hall.

Gould, W. A. 1990. *Glossary for the Food Industries.* Baltimore: CTI Publications.

Graf, E. 1991. *Food Product Development.* New York: Chapman and Hall.

Grenby, T. H. 1987. *Developments in Sweeteners.* Volume 3. New York: Chapman and Hall.

Grenby, T. H. 1989. *Progress in Sweeteners.* New York: Chapman and Hall.

Gump, B. H., and D. H. Pruett. 1993. *Beer and Wine Production: Analysis, Characterization, and Technological Advances.* Washington, DC: ACS.

Gunstone, F. D., et al. 1994. *The Lipid Handbook.* 2nd ed. New York: Chapman and Hall.

Hamilton, R. J. 1994. *Developments in Oils and Fats.* New York: Chapman and Hall.

Hamilton, R. J., and A. Bhati. 1987. *Recent Advances in Chemistry and Technology of Fats and Oils.* New York: Chapman and Hall.

Hardwick, W. A. 1994. *Handbook of Brewing.* New York: Marcel Dekker.

Haslam, E. 1989. *Plant Polyphenols: Vegetable Tanins Revisited.* New York: Cambridge University Press.

Hauschild, A. H. W., and K. L. Dodds. 1993. *Clostridium Botulinum: Ecology and Control in Foods.* New York: Marcel Dekker.

Heath, H., and G. A. Reineccus. 1986. *Flavor Chemistry and Technology.* New York: Chapman and Hall.

Hicks, D. 1990. *Production and Packaging of Non-Carbonated Fruit Juices and Fruit Beverages.* New York: Van Nostrand Reinhold.

Hill, M. J. 1988. *Nitrosamines Toxicology and Microbiology.* New York: VCH Publishers.

Hill, M. J. 1992. *Nitrate and Nitrite in Food and Water.* New York: Chapman and Hall.

Houghton, H. W. 1981. *Developments in Soft Drinks Technology.* Volume 2. New York: Chapman and Hall.

Houghton, H. W. 1984. *Developments in Soft Drinks Technology.* Volume 3. New York: Chapman and Hall.

Hubbert, W. T., and H. V. Hagstad. 1991. *Food Safety & Quality Assurance: Foods of Animal Origin.* Ames, IA: Iowa State University Press.

Hudson, B. J. F. 1990. *Food Antioxidants.* New York: Chapman and Hall.

Hui, Y. H. 1988. *United States Regulations for Processed Fruits and Vegetables.* New York: John Wiley & Sons.

Hui, Y. H. 1991. *Encyclopedia of Food Science and Technology.* Volumes 1–4. New York: John Wiley & Sons.

Hui, Y. H. 1991. *Data Sourcebook for Food Scientists and Technologists.* New York: John Wiley & Sons.

Hui, Y. H. 1993. *Dairy Science and Technology Handbook.* Volume I—Principles and Properties; Volume 2—Product Manufacturing: Volume 3—Applications Science, Technology, and Engineering. Deerfield Beach, FL: VCH Publishers.

Hume, U., and I. R. Hume. 1994. *Handbook of Industrial Seasonings.* New York: Chapman & Hall.

Hutchings, J. B. 1994. *Food Colour and Appearance.* New York: Chapman & Hall.

Ikan, R. 1991. *Natural Products: A Laboratory Guide.* 2nd ed. San Diego: Academic Press.

Imeson, A. 1992. *Thickening and Gelling Agents for Food.* New York: Chapman and Hall.

Ingram, D. K., et al. 1991. *The Potential for Nutritional Modulation of Aging Processes.* Trumbull, CT: Food and Nutrition Press.

Jackson, E. B. 1990. *Sugar Confectionery Manufacture.* New York: Chapman and Hall.

Jakubke, H. D. and H. Jeschkeit. 1993. *Concise Encyclopedia Chemistry.* Hawthorne, NY: Walter de Gruyter.

Jansen, G. R., et al. 1990. *Diet Evaluation: A Guide to Planning a Healthy Diet.* San Diego: Academic Press.

Jen, H. H. 1989. *Quality Factors of Fruits and Vegetables: Chemistry and Technology.* Washington, DC: American Chemical Society.

Johnson, I. T., and D. A. T. Southgate. 1994. *Dietary Fibre and Related Substances.* New York: Chapman & Hall.

Kader, A. A. 1992. *Postharvest Technology of Horticultural Crops.* 2nd ed. Davis, CA: ANR Publications.

Kanarek, R. B., and R. Marks-Kaufman. 1991. *Nutrition and Behavior: New Perspectives.* New York: Van Nostrand Reinhold.

Karmas, E., and R. S. Harris. 1988. *Nutritional Evaluation of Food Processing.* New York: Chapman and Hall.

Khan, R. 1993. *Low-Calorie Foods and Food Ingredients.* New York: Chapman and Hall.

Kinsella, J. E. 1992. *Advances in Food and Nutrition.* Volume 36. San Diego: Academic Press.

Kinsman, A. K., et al. 1994. *Muscle Foods.* New York: Chapman and Hall.

Kotsonis, F. N., and M. A. Mackey. 1994. *Nutrition in the '90s: Current Controversies and Analysis.* Volume 2. New York: Marcel Dekker.

Kritchevsky, D., and K. K. Carroll. 1994. *Nutrition and Disease Update: Heart Disease.* Champaign, IL: AOCS Press.

Krochta, J. M., et al. 1994. *Edible Coatings and Films To Improve Food Quality.* Lancaster, PA: Technomic Publishing Co.

Kulp, K. 1994. *Cookie Chemistry and Technology.* Manhattan, KS: American Institute of Baking.

Kulp, K., and R. Loewe. 1990. *Butters and Breadings in Food Processing.* St. Paul: American Association of Cereal Chemists.

Kurmann, J. A., et al. 1992. *Encyclopedia of Fermented Fresh Milk Products: An International Inventory of Fermented Milk, Cream, Buttermilk, Whey, and Related Products.* New York: Chapman and Hall.

Lanier, T. C., and C. M. Lee. 1992. *Surimi Technology.* New York: Marcel Dekker.

Larsson, K., and S. E. Friberg. 1990. *Food Emulsions, Revised and Expanded.* (Food Science and Technology Series, Volume 38). 2nd ed. New York: Marcel Dekker.

Lawrie, R. A. 1981. *Developments in Meat Science.* Volume 2. New York: Chapman and Hall.

Lawrie, R. A. 1984. *Developments in Meat Science.* Volume 3. New York: Chapman and Hall.

Lawrie, R. A. 1988. *Developments in Meat Science.* Volume 4. New York: Chapman and Hall.

Lawrie, R. A. 1991. *Developments in Meat Science.* Volume 5. New York: Chapman and Hall.

Lawson, H. 1995. *Food Oils and Fats.* 2nd ed. New York: Chapman and Hall.

Lees, R., and B. Jackson. 1973. *Sugar Confectionery and Chocolate Manufacture.* New York: Chapman and Hall.

Levenstein, H. 1993. *Paradox of Plenty: A Social History of Eating in Modern America.* New York: Oxford University Press.

Lorenz, K., and K. Kulp. 1991. *Handbook of Cereal Science and Technology.* (Food Science and Technology Series, Volume 41.) New York: Marcel Dekker.

Luh, B. S. 1991. *Rice.* Volume I—Production; Volume II—Utilization. 2nd ed. New York: Chapman and Hall.

Lusas, E. W. 1989. *Food Uses of Whole Oil and Protein Seeds.* Champaign, IL: AOCS.

Maarse, H. 1991. *Volatile Compounds in Foods and Beverages.* New York: Marcel Dekker.

MacKenzie, S. L., and D. C. Taylor. 1993. *Seed Oils for the Future.* Champaign, IL: AOCS.

MacRae, R., R. K. Robinson, and M. J. Sadler, eds. 1993. *Encyclopedia of Food Science, Food Technology, and Nutrition.* San Diego: Academic Press.

Manley, D. J. R. 1991. *Technology of Biscuits, Crackers and Cookies.* 2nd ed. New York: Chapman and Hall.

Marmion, D. M. 1991. *Handbook of U.S. Colorants: Foods, Drugs, Cosmetics, and Medical Devices.* 3rd ed. New York: John Wiley & Sons.

Marshall, W. E., and J. I. Wadsworth. 1994. *Rice Science and Technology.* New York: Marcel Dekker.

Martin, A. M. 1994. *Fisheries Processing Biotechnological Applications.* New York: Chapman and Hall.

Martin, R. E., and R. L. Collette. 1990. *Engineered Seafood Including Surimi.* Park Ridge, NJ: Noyes Data Corporation.

Martin, R. E., and G. J. Flick. 1990. *The Seafood Industry.* New York: Chapman and Hall.

Mathlouthi, M. 1994. *Food Packaging and Preservation.* New York: Chapman and Hall.

Matthews, R. H. 1989. *Legumes: Chemistry, Technology, and Human Nutrition.* (Food Science and Technology Series, Volume 47). New York: Marcel Dekker.

Matz, S. A. 1993. *Snack Food Technology.* 3rd ed. New York: Chapman and Hall.

Matz, S. A. 1991. *The Chemistry and Technology of Cereals as Food and Feed.* 2nd ed. New York: Chapman and Hall.

Matz, S. A. 1992. *Bakery Technology and Engineering.* 3rd ed. New York: Chapman and Hall.

Matz, S. A. 1992. *Cookie and Cracker Technology.* 3rd ed. New York: Chapman and Hall.

Mayer, D., and F. H. Kemper. 1991. *Acesulfame-K.* (Food Science and Technology Series, Volume 47). New York: Marcel Dekker.

Mazza, G., and E. Miniati. *Anthocyanins in Fruits, Vegetables, and Grains.* Boca Raton, FL: CRC Press.

Mennel, S., et al. 1992. *The Sociology of Food: Eating, Diet and Culture.* Thousand Oaks, CA: Sage Publications.

Mertz, E. T. 1992. *Quality Protein Maize.* St. Paul: American Association of Cereal Chemists.

Middledauf, R. D., and P. Shubik. 1989. *International Food Regulation Handbook: Policy, Science, and Law.* New York: Marcel Dekker.

Miller, Jones, J. 1992. *Food Safety.* St. Paul: Eagan Press.

Min, D. B., and T. H. Smouse. 1989. *Flavor Chemistry of Lipid Foods.* Champaign, IL: AOCS.

Minifie, B. W. 1989. *Chocolate, Cocoa, and Confectionery.* 3rd ed. New York: Chapman and Hall.

Mitchell, A. J. 1990. *Formulation and Production of Carbonated Soft Drinks.* New York: Chapman and Hall.

Moran, D. P. J., and K. K. Rajah. 1994. *Fats in Food Products.* New York: Chapman and Hall.

Mortan, S. 1993. *Food Irradiation: A Guidebook.* Lancaster, PA: Technomic Publishing.

Morton, I. D., and A. J. MacLeod. 1990. *Food Flavours. Part C: The Flavour of Fruits.* New York: Elsevier Science Publishers.

Nagy, S., et al. 1988. *Adulteration of Fruit Juice Beverages.* (Food Science and Technology Series, Volume 30). New York: Marcel Dekker.

Nagy, S., et al. 1993. *Fruit Juice Processing Technology.* Auburndale, FL: Agcience.

Nelson-Stafford, B. 1991. *From Kitchen to Consumer: The Entrepreneur's Guide to Commercial Food Production.* San Diego: Academic Press.

Nicolello, L. G., and J. Dinsdale. 1994. *Basic Pastrywork Techniques.* 2nd ed. New York: John Wiley & Sons.

Nielsen, S. S. 1994. *Introduction to the Chemical Analysis of Foods.* Boston: Jones & Bartlett Publishers.

Niewiadomski, H. *Rapeseed: Chemistry and Technology.* 1993. New York: Elsevier Science Publishers.

Nriagu, J. O. 1990. *Food Contamination from Environmental Sources.* New York: John Wiley & Sons.

O'Brien-Nabors, L., and R. Gelardi. 1991. *Alternative Sweetners.* (Food Science and Technology Series, Volume 48). 2nd ed. New York: Marcel Dekker.

Ockerman, H. W. 1989. *Sausage And Processed Meat Formulations.* New York: Chapman and Hall.

Ockerman, H. W. 1991. *Food Science Sourcebook, Parts 1 and 2.* 2nd ed. New York: Chapman and Hall.

Pennington, N., and C. W. Baker. 1991. *Sugar.* New York: Chapman and Hall.

Piggott, J. R. 1991. *Handbook of Sweeteners.* New York: Chapman and Hall.

Piggott, J. R., and A. Paterson. 1989. *Distilled Beverage Flavour: Recent Developments.* New York: VCH Publishers.

Piggott, J. R., and A. Paterson. 1994. *Understanding Natural Flavours.* New York: Chapman and Hall.

Pimentel, D., and C. W. Hall. 1989. *Food and Natural Resources.* San Diego: Academic Press.

Pomeranz, Y. 1991. *Functional Properties of Food Components.* 2nd ed. San Diego: Academic Press.

Pomeranze, Y., and C. E. Meloan. 1994. *Food Analysis: Theory and Practice.* 3rd ed. New York: Chapman & Hall.

Pronsky, Z. M. 1993. *Food Medication Interactions.* 8th ed. Pottstown, PA: Food Medication Interactions.

Rakosky, Jr., J. 1988. *Protein Additives in Foodservice Preparations.* New York: Van Nostrand Reinhold.

Reineccius, G. 1994. *Source Book of Flavors.* 2nd ed. New York: Chapman and Hall.

Rice, R. P., et al. 1990. *Fruit and Vegetable Production In Warm Climates.* Washington, DC: Scholium International.

Rodricks, J. V. 1992. *Calculated Risks: The Toxicity and Human Health Risks of Chemicals in our Environment.* Port Chester, NY: Cambridge University Press.

Ronsivalli, L. J., and E. R. Vieira. *Elementary Food Science.* 3rd ed. New York: Chapman and Hall.

Salunkhe, D. K., and S. S. Deshpande. 1991. *Foods of Plant Origin: Production, Technology, and Human Nutrition.* New York: Chapman and Hall.

Salunkhe, D. K., and S. S. Kadam. 1989. *Handbook of World Food Legumes: Nutritional Chemistry, Processing Technology, and Utilization.* Boca Raton, FL: CRC Press.

Salunkhe, D. K., et al. 1990. *Dietary Tannins: Consequences and Remedies.* Boca Raton, FL: CRC Press.

Salunkhe, D. K., et al. 1991. *Potato: Production, Processing, and Products.* Boca Raton, FL: CRC Press.

Salunkhe, D. K., et al. 1991. *Storage, Processing and Nutritional Qaulity of Fruits and Vegetables.* Volume II—Processed Fruits and Vegetables. 2nd ed. Boca Raton, FL: CRC Press.

Salunkhe, D. K., et al. 1992. *World Oilseeds: Chemistry, Technology, and Utilization.* New York: Chapman and Hall.

Santerre, C. R. 1994. *Pecan Technology.* New York: Chapman & Hall.

Saxby, M. 1993. *Food Taints and Off-Flavours.* New York: Chapman and Hall.

Shahidi, F. 1994. *Flavour of Meat and Meat Products.* New York: Chapman and Hall.

Shibamoto, T., and L. F. Bjeldanes. 1993. *Introduction to Food Toxicology.* San Diego, CA: Academic Press.

Sikorski, Z. E. 1993. *Seafood: Resources, Nutritional composition, and Preservation.* Boca Raton, FL: CRC Press.

Sikorski, Z. E. 1994. *Seafood Proteins.* New York: Chapman & Hall.

Smith, J. 1991. *Food Additive User's Handbook.* New York: Chapman and Hall.

Smith, J. 1993. *Technology of Reduced-Additive Foods.* New York: Chapman and Hall.

Stauffer, C. E. 1991. *Functional Additives for Bakery Foods.* New York: Chapman and Hall.

Stear, C. A. 1990. *Handbook of Breadmaking Technology.* New York: Chapman and Hall.

Stegink, L. D., and L. J. Filer Jr. 1984. *Aspartame: Physiology and Biochemistry.* New York: Marcel Dekker.

Szuhai, B. F. 1989. *Lecithins: Sources, Manufacture & Uses.* Champaign, IL: American Oil Chemist's Society.

Tainter, D. R., and A. T. Grenis. 1993. *Spices and Seasonings: A Food Technology Handbook.* Deerfield Beach, FL: VCH Publishers.

Talburt, W. F., and O. Smith. 1987. *Potato Processing.* 4th ed. New York: Chapman and Hall.

Taylor, S. L., and R. A. Scanlan. 1989. *Food Toxicology: A Perspective on the Relative Risks.* New York: Marcel Dekker.

Thorne, S. 1989. *Developments in Food Preservation.* Volume 5. New York: Chapman and Hall.

Ting, S. V., and R. Rouseff. 1986. *Citrus Fruits and Their Products: Analysis and Technology.* New York: Marcel Dekker.

Underriner, E., and I. Hume. 1994. *Handbook of Industrial Seasonings.* New York: Chapman and Hall.

Vail, R. 1994. *Food Safety and Quality Assurance.* New York: Marcel Dekker.

Varnam, A., and J. Sutherland. 1994. *Milk and Milk Products Technology, Chemistry and Microbiology.* New York: Chapman and Hall.

Verzele, M., and D. de Keukeleire. 1991. *Chemistry and Analysis of Hop and Beer Bitter Acids.* New York: Elsevier Science Publishers.

Vetter, J. L. 1992. *Food Labeling: Requirements for Bakery and Related Products.* Manhattan, KS: American Institute of Baking.

Volker, B. 1988. *Vademecum for Vitamin Formulations.* Boca Raton, FL: CRC Press.

Wade, P. 1988. *Biscuits, Cookies and Crackers.* Volume 1—The Principles of the Craft. New York: Chapman and Hall.

Walter, D. E., et al. 1991. *Sweeteners: Discovery, Molecular Design and Chemoreception.* Washington, DC: American Chemical Society.

Walter, R. H. 1991. *The Chemistry and Technology of Pectin.* San Diego, CA: Academic Press.

Weichmann, J. 1987. *Post-Harvest Physiology of Vegetables.* New York: Marcel Dekker.

Whistler, R. L., and J. N. BeMiller. 1992. *Industrial Gums: Polysaccharides and Their Derivatives.* San Diego: Academic Press.

Whiteley, P. R. 1971. *Biscuit Manufacture.* New York: Chapman and Hall.

Woteki, C. E., and P. R. Thomas. 1992. *Eat for Life: The Food and Nutrition Board's Guide to Reducing Your Risk of Chronic Disease.* Washington, DC: National Academy Press.

Yalpani, M. 1988. *Polysaccharides: Syntheses Modifications and Structure/Property Relations.* New York: Elsevier Science Publishers, Science & Technology.

Young, J. H. 1989. *Pure Food: Securing the Federal Food and Drugs Act of 1906.* Princeton, NJ: University Press.

Zeuthen, P., et al. 1990. *Processing and Quality of Foods Volume 1: High Temperature/Short Time (HTST) Processing Guarantee for High Quality Food with Long Shelf Life.* New York: Chapman and Hall.

Zeuthen, P., et al. 1990. *Processing and Quality of Foods Volume 2: Food Biotechnology: Avenues to Health and Nutritious Products.* New York: Chapman and Hall.

Zeuthen, P., et al. 1990. *Processing and Quality of Foods Volume 3: Chilled Foods: The Revolution in Freshness.* New York: Chapman and Hall.